FISH FERMENTATION

FISH FERMENTATION

by
Jayesh Agarwal | Anubhav Amble | Ojas Bhandari

Wordpen Academics
Publishing Science That Matter

ISBN: 978-93-6084-285-7 (Hardback)

Published by : **WORDPEN ACADEMICS**
4736/23, Ansari Road, Darya Ganj,
New Delhi – 110 002
Phone: +91-011-35004336
E-mail: info@wordpenacademics.com

Printed at **:** **Replika Press Pvt. Ltd.**

Exclusive Distributor: ***Bio-Green Books, New Delhi***

Preface

In the vast tapestry of culinary traditions around the world, fish fermentation stands as a time-honored practice that has been passed down through generations, enriching cuisines and cultures with unique flavors and textures. From the pungent fish sauces of Southeast Asia to the savory fish pastes of Scandinavia, fermented fish products have long been cherished for their distinct taste profiles and culinary versatility. As we explore the art and science of fish fermentation, we embark on a journey of discovery that unveils the rich history, cultural significance, and culinary innovations of this ancient preservation technique.

This book, titled "Fish Fermentation," is a celebration of the diverse traditions and techniques of fermenting fish across different cultures and regions. Drawing upon the expertise of food scientists, chefs, anthropologists, and fermentation enthusiasts, this volume offers insights, recipes, and cultural perspectives on the fascinating world of fermented fish products.

From the microbial processes that drive fermentation to the sensory attributes that define fermented fish flavors, "Fish Fermentation" delves into the science behind this age-old preservation method. Through a blend of scientific explanations, historical anecdotes, and culinary anecdotes, readers will gain a deeper understanding of the principles and techniques of fish fermentation, as well as the cultural significance of fermented fish products in various culinary traditions.

Moreover, this book showcases the diverse array of fermented fish products found around the world, from Korean kimchi to Icelandic hakarl. By exploring traditional recipes, modern adaptations, and innovative uses of fermented fish, we aim to inspire readers to experiment with fermentation in their own kitchens and appreciate the rich tapestry of flavors that fermented fish products have to offer.

As we embark on this journey of exploration and discovery, it is our hope that "Fish Fermentation" will serve as a source of inspiration, knowledge, and culinary delight for food enthusiasts, chefs, and anyone curious about the art of fermentation. Whether you are a seasoned fermenter seeking new recipes or a novice eager to explore the world of fermented foods, this book invites you to immerse yourself in the ancient tradition of fish fermentation and discover the flavors of cultures past and present.

Jayesh Agarwal | Anubhav Amble | Ojas Bhandari

Contents

1

INTRODUCTION

Fermented fish is a traditional preservation of fish. Before refrigeration, canning and other modern preservation techniques became available, fermenting was an important preservation method. Fish rapidly spoils, or goes rotten, unless some method is applied to stop the bacteria that produce the spoilage. Fermentation is a method which attacks the ability of microbials to spoil fish. It does this by making the fish muscle more acidic; bacteria usually cease multiplying when the pH drops below 4.5.

A modern approach, biopreservation, adds lactic acid bacteria to the fish to be fermented. This produces active antimicrobials such as lactic and acetic acid, hydrogen peroxide, and peptide bacteriocins. It can also produce the antimicrobial nisin, a particularly effective preservative.

Fermented fish preparations can be notable for their putrid smell. These days there are many other techniques of preserving fish, but fish is still fermented because some people enjoy the taste.

Alaska has witnessed a steady increase of cases of botulism since 1985. It has more cases of foodborne botulism than any other state in the United States of America. This is caused by the traditional Eskimo practice of allowing animal products such as whole fish, fish heads, walrus, sea lion, and whale flippers, beaver tails, seal oil, birds, etc., to ferment for an extended period of time before being consumed. The risk is exacerbated when a plastic container is used for this purpose instead of the old-fashioned, traditional method, a grass-lined hole, as the botulinum bacteria thrive in the anaerobic conditions created by the air-tight enclosure in plastic.

Fermentation is the microbial process of transforming food through the use of bacteria or fungus, as well as the enzymes produced by these microorganisms. This transformative ability of microbes has been utilized by people for production of fermented beverages and foods in order to preserve food and make the food more digestible with reduced harmfulness. Fermented foods involve controlling the microbial growth, which helps in the biotransformation of bioactive phenolics and volatile compounds into new edible forms with enhanced flavors and aromas. The microbial population in fermented food products deriving indigenously or from external 'starter cultures' causes chemical and textural changes to the food The microbial population in food can either cause spoilage of the food or produce a food product that is edible and safe for food, or there is a very thin line, determined by taste and flavor, between putrid and fermented food wherein the first one is undesirable, but the latter becomes a delicacy. Fermented food has evolved with the development of human civilization since ancient times, partly sustaining the human population in the past and providing nutrition (calories, proteins, vitamins, minerals) at a reasonable cost. In particular, fermented fish is one of the most popular among the many different varieties of fermented food products. Although fermented food naturally contains live microorganisms (probiotics), the health benefits of their consumption are still uncertain.

The usage of fermented fish as a dish in traditional culinary habits dates all the way back to ancient times in many societies throughout the world. The earliest fermented fish sauce ever recorded is garum in classical Rome, which was prepared from the viscera and blood of mackerel. Consigned to academic and cultural invisibility in food studies for the most part, this fish sauce has only recently come to be seen as having a striking similarity with fish sauces in Southeast Asia, from its umami taste and flavor and its ability to radically transform everyday dishes (Fig. 1) to its being cheap and commonly available. In many cultures throughout history, people have developed diverse means to preserve food resources like fish during periods of scarcity following an abundant seasonal supply of fish. The fish had to be preserved via variety of ways due to its perishable nature, including drying, salting, and fermenting.

Apart from East, Southeast and South Asia (Fig. 2), countries in other parts of the world like the Middle East, Africa and the Scandinavian countries in Europe also produce and consume fermented fish products, like surstromming (fermented herring) and rakorret (fermented trout) in Scandinavia and fermented anchovy in France and Spain. There is a considerable degree of artisanship associated with fermenting fish in all the communities where it is a part of their everyday life and culture. In Africa, there are three basic techniques, namely fermentation with salting and drying, fermentation and drying without salting, and fermentation with salting but no drying. Moreover, while fermented fish pastes and fish sauces are more popular in Southeast Asia, fermented fish products in Africa are usually presented in the form of the whole fish or large chunks cut from it. Again, Asian and African cultures mostly use the traditional methods of (spontaneous, i.e., without the addition of starter cultures) fermentation unlike those in Europe, North America, Australia, and New Zealand where the fermented food products are normally made with specific starter cultures, i.e., microbial inocula which are used to initiate fermentation.The inoculum used as a starter culture may be a single strain of bacteria or a consortium isolated from natural fermented food products.

Strategies such as limited drying, smoking, fermentation, and strategically utilizing substrates to stimulate the development of some beneficial microorganisms that inhibit the growth of pathogenic microbes are used to produce fish products in a traditional manner. The fermented fish products are a significant source of protein, forming an integral part of staple diet, and possessing great cultural significance in many Asian countries such as Korea, Japan, China, Thailand, Cambodia, Malaysia, Philippines, and Indonesia and in Asian foodways in general. The term 'foodways' denotes the composite set of culinary practices, customs, beliefs, folklore, the intersection of ethnicity and food habits, material aspects of food, memories and nostalgia regarding food and culture, and so on, that defines the connection between a particular culture and its food habits and cuisine. In India, ranging from the Northeast to the South, it is an integral part of food culture. Other South Asian countries like Sri Lanka, Bangladesh, Bhutan, and Nepal also have their own fermented fish products. Fish fermentation is a small-scale industry and can provide direct employment to the people involved in it with very simple artisanal equipment. According to a report by the FAO, the market and production of fermented products in a country are influenced by the food habits of the people and market demands. In the developing countries of Southeast Asia, South Asia and Africa, there is a good market for fermented fish products.

In India, fermented fish is popular and various types of fermented fish products have been reported from its Northeastern region. Matkashidal is a widely known fermented fish that is economically viable in the Northeast Indian states of Manipur and Tripura. Fermented fish is a staple food in South and Southeast Asian foodways, and the product has a large consumer base in the region. However, these products are typically produced at the family or village level using traditional methods in the region. Some products like fish sauce and shrimp paste are, however, produced on an industrial scale and have good market potential. This has led to some nations embarking on improving the quality of such products, like the Cambodian government, which introduced strict industry standards (underlining certain specific characteristics like color and smell) in 2016 for the production of prahok, a type of fermented fish pastes and a national staple food. This was due to the realization that the popularity of prahok among both resident and diasporic Cambodians made it necessary to monitor the health and safety issues related to the product.

In their study of the cultural contexts and significance of fermented fish and fish products in Southeast Asia, Ruddle and Ishige argued that this region should be considered a central point of origin of fermented fish products, as the freshwater fishes in the rice producing region (particularly in the rice fields) began to be fermented by the earliest indigenous populations. The primary reasons for the fermentation of fish in this region since ancient times are the seasonal availability of fish and its scarcity during the long dry months. The abundance of water bodies and the fact that the cultivation of rice required a huge and consistent supply of water during its growing period (something that the region's climate ensured) created a conducive environment for fish to flourish during the rainy season. Furthermore, archaeological excavations and studies reveal that Southeast Asia was a hub of salt production (apart from more well-known archaeological sites in China) since prehistoric times. These circumstances together compelled and facilitated the preservation of fish using fermentation methods, given the high degree of humidity in the region , leading to fermented fish and fish products dominating the food cultures in the region. One notable example is Northeastern Thailand where rice and fish are staple foods, while salt is the most commonly available preservative. As a result, these resources consequently define the food cultures in this region, in the sense that fish fermented using salt as an additive is a primary side dish that invariably accompanies and renders the staple diet of rice tasteful. Due to the relative lack of economic security and modernized ways of life in many parts of Southeast Asia (especially in the rural areas like in Northeastern Thailand), producing and consuming fermented fish were (and still are) a cost-effective and accessible way of including proteins and minerals like calcium and phosphorus in the everyday diet. In addition, fermented fish, with its umami taste, relieved the mundanity of the rice and vegetable diet of the local population.

In the case of India's Northeast, the climate, with its heavy rainfall and high humidity (making this region the wettest in India, and indeed, among the wettest regions in the world), made it impossible to dry the fish, while the sun shone, simply because there was not enough sunlight, unlike in the drier parts of India. The monsoon season from May to September naturally coincided with the fishing season, followed by a long period of scarcity of fresh fish in the region. This necessitated an efficient and relatively faster method of fish preservation through fermenting, and it is thus that this part of India has a central presence in the fermented food map of India.

Another important factor is the similarity of the Northeast Indian communities and their ways of life with the larger Southeast Asian region and cultures. This owes partly to the strategic geographical location of the region, connecting, on the one hand, the Indian subcontinent with Southeast Asia. This contiguity as well as continuity extends to the shared food habits of the two regions (Northeast India and Southeast Asia) characterized by a centrality of rice and fish in Southeast Asian and Northeast Indian foodways. These shared sensibility and shared food preferences, again, embody a continuity of cultural habits and ways of life going back to ancient times that are also reflected, for example, through similarities in dress and clothing, as well as music and musical instruments (among other aspects). In fact, archaeological excavations have revealed that the Neolithic culture in Northeast India produced cord-marked pottery that is associated with the Neolithic cultures in Southeast Asia. When placed within this framework, the cultural and gastronomical centrality of fermented fish does not merely symbolize the shared cultural identities of the various communities in the two regions but also points to the fact that foodways create a unique cultural map that is extremely fluid and extends across political boundaries.

Given their high cultural and economic significance, the study of fermented fish products gives an ample scope for further research. For these reasons, this review is organized to summarize the unique fermented fish products from Asia, their nutritional composition, cultural significance, and health implications. Furthermore, this article has focused on all Asian fermented fish cuisines. Because of the abundance of the literature on traditional Chinese cuisine, numerous examples of Chinese fermented food are readily available and recognized around the world. The database, however, contains only a limited amount of information about other Asian countries. Due to a lack of fermented fish composition analysis in the rest of Asia, we focused on fermented fish from South and Southeast Asia to narrow the information gap.

FERMENTATION

The fermentation is the conversion of a carbohydrate such as sugar into an acid or an alcohol. More specifically, fermentation can refer to the use of yeast to change sugar into alcohol or the use of bacteria to create lactic acid in certain meals. Fermentation occurs naturally in many different meals given the good situation, and humans have intentionally made use of it for many thousands of years.

The earliest uses of fermentation were most likely to create alcoholic beverages such as mead, wine, and beer. These other drink liquid not water may have been created as far back as 7,000 BCE in parts of the Middle East. The fermentation of meals such as milk and different vegetables probably happened sometime a few thousand years later, in both the Middle East and China. While the general principle of fermentation is the same across all of these drinks and foods, the precise methods of achieving it, and the end results, differ.

Beer is made by taking a grain, such as barley, wheat, or rye, germinating and drying it, and pupping it into a mash. This mash is then mixed with hot water, and some fermentation begins. After being further treated, the liquid is transferred to a fermentation vessel, where yeast is included to the mixture. This yeast "eats" the sugar present in the mash and converts it into carbon dioxide and alcohol. After a few weeks of fermentation and a further period of conditioning, the beer is ready to be filtered and consumed.

Wine is made by using different factors like: sugar, rich juices and fermentation is include different- different liquid. Grapes are crushed to release the sugar-rich juices, which are then either transferred quickly away from the skins or left to rest for a time to absorb some of the flavour, tannings, and colour of the skins. Yeast is included, and the grape juice is allowed to ferment for a number of weeks, at which point it is moved to different containers and fermented at a slower rate, and eventually aged or bottled.

Pickling meals, such as cucumbers, may be accomplished by submerging the vegetable one wants to pickle in a salty water solution with vinegar included. Over time, bacteria create the lactic acid that gives the meal its distinctive flavour and helps to preserve it. Other meals can be pickled simply by packing them in dry salt and allowing a natural fermentation process to occur.

Milk can also be cultured, and people have been using fermentation with dairy products for nearly 5,000 years. It is speculated that early fermented dairy, such as yoghourt, was the result of a natural process of fermentation that occurred when the milk was cultured by bacteria that dwelt in skin sacks used to store dairy. Yoghourt these days is made by including a number of special bacteria, such as *L.* acidophilus and *L. bulgaricus* to milk and keeping it at the proper temperature. The bacteria begin converting the sugar in the dairy to lactic acid, eventually creating what we know as yoghourt.

Area of Biotechnology

There are some biotechnology are given below:-

(i) Industrial Production

Industrial Production: Traditionally, industrial ethanol is manufactured via the acid catalysed hydration of ethylene. One synthetic route of alcohol by hydration of ethene involves the use of concentrated sulfuric acid (H2SO4). Unfortunately the use of sulfuric acid complicates the process due to concerns about safety, corrosion, and the environment. Presently many technology produced alcohol and silica aero gels impregnated with phosphoric acid and determine the many alcohol which are related and included. The distinct feature of this process are that the reaction can be a one stage process, the catalyst is regenerated, and concerns about safety, corrosion, and the environment are diminished. The reaction with phosphoric acid is as follows: This method has been chosen due to the relative low cost of the ethylene.

$$H2C = CH2 + H2PO3 + H2O \longrightarrow HOCH2CH3 + H2\ PO3$$

One problem with the industrial production and use cell available resources of ethanol is that ethanol is the alcohol found in beverages. Thus, industrial alcohol could be the source of large amounts of wine, beer and other alcohol not use of water. But it has been the practice to tax and control alcohol for beverages and industrial ethanol could be the source for a large amount of illegal alcohol.

In order to reduce the need for strict control and heavy taxation on industrially produced ethanol, the alcohol is denatured. Denaturing is a process of including other compounds to the ethanol to render it unfit for consumption. Denaturants are selected to give the ethanol a disagreeable taste or odor and in some cases a distinctive colour. In some cases the substances included are toxic and produce gastric disturbances upon ingestion and/or other unpleasant symptoms. A large number of different "denaturants" are utilized dependent upon the use for which the ethanol is intended. These denaturants include methyl isobutyl ketone, pyronate, kerosene, acetone, turpentine, amyl alcohol, methyl alcohol, and various butyl alcohols. In some cases more than one denaturant is utilized.

Vinegar was originally produced by terms learning wine in shallow bowls or partially filled barrels where it was slowly oxidised to vinegar by the development of a natural flora. The problem was minimized by restricting the initial wart concentration such that the growth of the cells limited by the available of the carbon sources rather than oxygen.

Industrially produced ethanol has many uses adding in solvent based paints, pharmaceuticals, perfumes, cleaning products for home and car, lacquers, and inks.

Further ethanol and other alcohols have been added to gasoline to produce alcohol-containing fuels. Since ethanol contains particular partially oxidised carbon atom, it is flammable and can be used as a fuel. The usual ratio of ethanol to gasoline is ninety parts gasoline to 10 parts ethanol while that for methanol is 97 parts gasoline to 3 parts methanol. Some cars have been designed to run on pure methanol. Gasohol has higher antiknock properties (higher octane) than gasoline, burns more slowly, coolly, and completely. Above all, it is more expensive and power intensive to produce and may produce environmental hazards to the individual and to the atmosphere.

Fermentation and Industrial and Beverage Production: All wine and beer, other alcohol and much of that used in industry is formed through fermentation of a type of products adding grain such as corn, potato mashes, fruit juices, and beet and cane sugar molasses. (In earlier years, until about 1947, the largest proportion of the production of industrial alcohol was from fermentation, but as was described above the hydration of ethene now provides the biggest source of industrial alcohol). Fermentation can be defined as an enzymatically anaerobic controlled transformation of an organic compound. With respect to alcohol, we are referring to the conversion of sugars to ethanol by microscopic yeasts in the absence of oxygen. The equation for the fermentation of glucose is:

C6H1206 —>in the presence of yeast 2CH3CH2OH + 2CO2

Biochemistry have different-different symbolic notation in this figure. It shows step to step transfer of glucose to ethanol through intermediates, pyruvate and acetaldehyde.

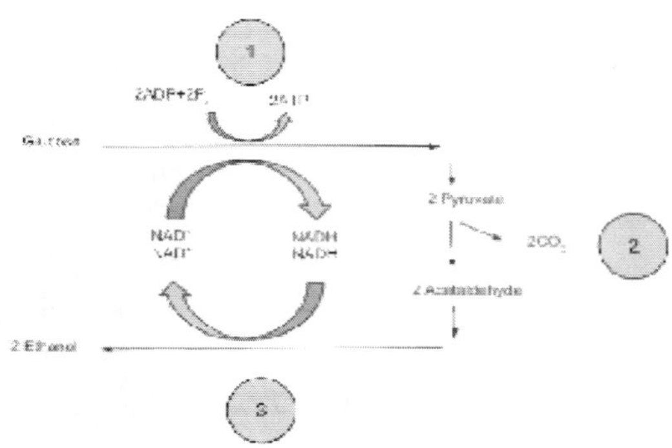

The fermentation included contain near about 3 to 5 per cent ethanol in beer and 12 to 15 per cent ethanol in wine and sherry, because ethanol can not achieved by the fermentation.

Distillation is a process that uses differences in boiling points to separate compounds. In the case of alcohol and particularly ethanol, knowledge that the boiling point of pure water is 100C, while that of ethanol is 78.3C allows the separation of the ethanol from the water by adjusting the distillation temperature to a point higher than that for ethanol, but lower than that for water. Thus, the concentration of ethanol can be enhanced by removing it as a distillate from the ethanol-water solution. Spirits such as gin, scotch, bourbon, and vodka, as well as liqueurs, cordials, and bitters are examples of beverages made from distillation. The distillation procedure also allows for the concentration of components of the beverage which provide some distinctive flavour.

The fermentation included contains near about 3 to 5per cent ethanol in beer and 12 to 15per cent ethanol in wine and sherry because ethanol cannot active by the fermentation.

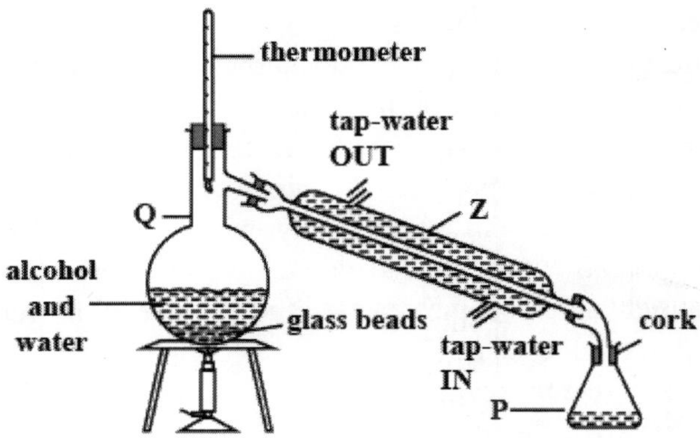

Pure ethanol cannot be obtained via conventional distillation of a water-ethanol mixture because a constant boiling mixture forms consisting of 95per cent ethanol-5 per cent water. Such a mixture is referred to as an azeotrope azeotropic = a liquid mixture that is characterized by a constant concentration and constant minimum or maximum boiling point which is lower or higher than any of the components. Vacuum distillation or addition of another substance to the mixture. Often times the compound added is highly toxic such as benzene, therefore] absolute alcohol must never be consumed.

The amount of ethyl alcohol in liquid that wine and beer. Thus, there are differences in the amount of alcohol between beer, wine, champagne and distilled spirits. The amount of alcohol is given as a percentage and also in "proof". The proof of an alcohol beverage is equal to twice the percentage of ethyl alcohol contained therein. Thus, 100 proof ethanol is 50 per cent and 50 proof ethanol is 25 per cent, etc.

Magou Fermentation

The Magou is very famous in South Africa, basically Bantu people linked very much and Magou is a lactic acid fermented porridge made from maize. To make magua a 10 per cent maize meal slurry is cooked, cooled and inoculated with wheat flour, which contains the bacteria. Magou is also produced on industrial scale and is then packed in cartons. In the industrial process the magou is inoculated with lactobacillus delbreuckii cultures.

Kefir Fermentation

Kefir fermentation is similar to yoghourt fermentation. Yoghourt is only fermented by bacteria but kefir fermentation involves the help of bacteria as well as yeasts. These yeast produce some alcohol and Co2 which gives kefir its typical fizzy aspect. Kefir is inoculated with special kefir grains. These grains are mixtures of bacteria and yeasts in a matrix of proteins, lipids and carbohydrates. Kefir fermentation is done at room temperature, which makes the process easier. On the other hand, not everyone likes the taste of kefir.

ACETIC ACID FERMENTATION

The acetic acid fermentation is control by the *Acetobacter aceti* and the acetic acid is very energetic liquid which produce inhibits by the micro-organism. To alleviate this inhibitory effect, we developed an electrodialysis fermentation method such that acetic acid is continuously removed from the broth. The fermentation unit has a computerised system for the control of the pH and the concentration of ethanol in the fermentation broth. The electrodialysis fermentation system resulted in improved cell growth and higher productivity over an extended period; the productivity exceeded that from non-pH-controlled fermentation. During electrodialysis fermentation in our system, 97.6 g of acetic acid was produced from 86.0 g of ethanol; the amount of acetic acid was about 2.4 times greater than that produced by non-pH-controlled fermentation (40.1 g of acetic acid produced from 33.8 g of ethanol). Maximum productivity of electrodialysis fermentation in our system was 2.13 g/h, a rate which was 1.35 times higher than that of non-pH-controlled fermentation (1.58 g/h).

BUTYRIC ACID FERMENTATION

The fermentation butyric acid is also found as a hexyl ester in the oil of Heracleum giganteum and as an octyl ester in parsnip it has also been noticed in the fluors of the flesh and in perspiration.

Scope of Fermentation

The Fermentation programme communicate knowledge about the main terms of fermentation technology in the production of food like that beer wine and other liquid and pharmaceuticals It includes some information on historical aspects of biotechnology and also about laboratory research and development.

This Unit has a modular structure, offering teachers the opportunity to select items for a particular teaching situations. It is recommended that the topic is introduced within a local and topical context, to awaken the interest of the students.

Industrial microbiology

The Industrial microbiology is use the that is make the product of food and other use of industrial products. micro-organisms in the manufacture of food or industrial products. The use of micro-organisms for the production of food, either human or animal, is often considered a branch of food microbiology. The micro-organisms used in industrial processes may be natural isolates, laboratory selected mutants or genetically engineered organisms.

FOOD MICROBIOLOGY

Yoghourt, cheese, chocolate, butter, pickles, sauerkraut, soya sauce, vitamins, amino acids, food thickeners, alcohol, sausages, and silage are all produced by industrial microbiology processes. 'Good' bacteria such as probiotics are becoming increasingly important in the food industry. Today the food industry is increasing the work in the international market and also provide the batter facilities.

BIOPOLYMERS

A huge is the variety of biopolymers, as polysaccharides, polyesters, and polyamides, are produced by micro-organisms. These products range from viscous solutions to plastics. The genetic manipulation of micro-organisms has permitted the biotechnological production of biopolymers with tailored material properties suitable for high-value medical application such as tissue engineering and drug delivery. Industrial microbiology can be used for the biosynthesis of xanthan, alginate, cellulose, cyanophycin, poly (gamma-glutamic acid), levan, hyaluronic acid, organic acids, oligosaccharides and polysaccharides, and polyhydroxyalkanoates by jose landron.

BIOREMEDIATION

The Biodegradation of pollutants could be used to cleanup contaminated environments. These bioremediation and biotransformation methods harness naturally occurring microbes to degrade, transform or accumulate a huge range of compounds adding hydrocarbons polychlorinated biphenyls (PCBs), polyaromatic hydrocarbons (PAHs), pharmaceutical substances, radionuclides and metals.

WASTE BIOTREATMENT

By modern society is generated the quantities of wastes which is released to the microorganisms. Biotreatment, the processing of wastes using living organisms, is an environmentally friendly, relatively simple and cost-effective alternative to physico-chemical clean-up options. Confined environments, such as bioreactors can be employed in biotreatment processes.

Wastewater Treatment

The wastewater treatment is one of the most important biotechnological processes, Biotechnological processes have many process to related the waste water treatment. Which have been used for over a century to treat municipal and industrial wastewaters. Culture-independent molecular techniques have been used to study the diversity and physiology of ecologically important microorganisms in wastewater treatment processes.

A number of new exciting insights into the structure, function, and dynamics of complex microbial communities in wastewater treatment processes have been gained, which have significantly expanded our understanding of process design, operation and control. Microbes play a vital role in the cycling of nitrogen in wastewater treatment processes (including anaerobic ammonia oxidation processes) and methane fermentation processes. And also including the anaerobic ammonia oxidation process which process are helpful for the waste water treatment.

HEALTH CARE AND MEDICINE

Micro-organisms are used to produce human or animal biological such as insulin, growth hormone, and antibodies. Diagnostic assays that use monoclonal antibody, DNA probe technology or real-time PCR are used as rapid tests for pathogenic organisms in the clinical laborarory.

Micro-organisms may also help in the treatment of diseases such as cancer. Research shows that clostridia can selectively target cancer cells. Various strains of non-pathogenic clostridia have been shown to infiltrate and replicate within solid tumours. Clostridia so have the potential to deliver therapeutic proteins to tumours. Lactobacillus spp., and other lactic acid bacteria possess numerous potential therapeutic properties adding anti-inflammatory and anti-cancer activities.

ARCHAEA

Examination of microbes living in unusual environments lead to discovery of microbes with new abilities that can be harnessed for industrial purposes. For e.g. The high temperatures, slat, low PH or temperature, high radiation.

CORYNEBACTERIA

Corynebacteria are various type of gram-positive bacteria establish in range of diverse ecological niches as well as (1) soil (2) vegetable (3) skin (4) sewage. Corynebacterium glutamicum is of immense industrial momentous and is one of the biotechnologically most important bacterial species with an annual production of more than two million tons of amino acids, mainly L-glutamate and L-lysine. The genome sequence of C. glutamicum has been published.

XANTHOMONAS

The Xanthomonas consists of 20 plant-associated species, many of which cause moments illness of crops and other plants particular species comprise multiple pathovars, characterized by distinctive host specificity or mode of infection. Bacteria of the genus Xanthomonas are able to produce the acidic exopolysaccharide xanthan gum. Because of its physical properties, it is widely used as a viscosifer, thickener, emulsifier or stabiliser in both food and non-food industries.

Production of Microbial Cell

The machines and human workers both are control the production because the products were make by the machine and human in the separate area. So we can say that the workers would worked on a production leave. In cell production, or cellular manufacturing workers are organized into multi-skilled teams. Each team is responsible for a particular part of the production process including quality control and health and safety. Each cell is made up of one team who deliver finished items on to the next cell in the production process. UPS and can lead to efficiency improvements due to increased motivation (team spirit and added responsibility given to cells) and workers sharing their skills and expertise.

Economics of Fermentation

The expanding our operation are related to the contact between commence and craft in the food production. Usually I make soda in 5-gallon batches. The process is fairly time-consuming, but it fits in well with other chores and there is no obligation to brew a certain amount at a certain time. Since I enjoy brewing several times a week, I produce a surplus—far more than our own family can drink. To expand to a commercial level, though, would mean changes in the way I brew, because as it stands I can only net about $20 per hour of labour. To be commercially viable, I would need to exploit efficiencies of scale by buying better equipment: a bottle-washing machine, bottler, larger fermentation vessels, etc. Then it is no longer a kitchen hobby; it is a business that must consider shipping, legal licensing, labeling laws, sanitation regulations, accounting, etc.

THE COMPROMISES OF COMMERCE

Some people are naturally inclined towards business. There are more and more small food crafting businesses these days, To pay a premium for their products. But it is more than a matter of hobby vs business—there are certain compromises one must make to bring production past a certain critical volume. The critical volume for fermented foods is especially low, because the product is alive and working. Lactofermented soda keeps fermenting in the bottle, for instance, leading to foaming and spraying when you open it, or even dangerous exploding bottles, There is a good reason that mass-marketed soft drinks are dead. In fact it is a necessity in the context of national brands, centralised production, and mass distribution. The food is produced to use of different way and determine process and dimandsinthe way its distributed and the products are sold.

The most compromise would have to manufacture to sell fermented soda on economically, this is constant refngeration warming labels and other security terms used on the production of soda fermentation or plastic bottles.

Similar compromises apply to most of the fermented foods that have survived the last century of food industrialisation. Pickles and relish are no longer fermented at all, but preserved in vinegar and sterilised with heat in the canning process. Wine is treated with sodium metabisulfite before fermentation to destroy wild bacteria and froth appearing in fermentation that make the results less predictable. Beer is usually pasteurised or microfiltered to kill or remove living preparation used for raising dough. Yogurt survives, but it just isn't as best after the first day; the same is true of bread. Sauerkraut is usually pasteurised. To be sure there are niche brands, available in health-food stores, which are still living foods, but then freezing or refrigeration is necessary. This is rather ironic, since a major motivation for fermenting foods in the first place was to preserve them, in the days before refrigeration.

CLASSIFICATION OF FERMENTATION

Money Reigns Supreme

The market economics is not applying lows of market because money may be facilitate replace various friends and other person. But essence money is anonymous for resource of energy that is universal medium of exchange. You don't seek to maximize profit. You don't raise your prices to the maximum just because you can. You are not doing it for the money; you are doing it for your family and for the neighbours. In an economy of reciprocation and social exchange; that is, in an economy that is not primarily a money economy, "economic efficiency" takes on a different meaning.

The anonymous the customer are stands more money to sole. In today's multi-level, automated, and standardised meal production and distribution system, the consumer is almost totally anonymous to the farmer, the commodity buyer, the processing factory, and even the grocer. There is no reason to care about the wholesomeness of the product, except to the extent necessary to conform to whatever regulations are enforced, and whatever the public might find out about. No reason? Oh pardon me, I forgot about altruism. Yes of course, a company might make products better than they need to be out of a abstract altruism, but when the pressures of market competition come to bear, such altruism quickly degenerates into sloganeering and PR. Some version of "caring about the health of the consumer" surely appears in the mission statements of all the major meal corporations, adding the most egregious violators of the public trust. In other words, it is hard to genuinely care about someone you don't even know. Compassion in the abstract is almost always a self-deception.

Much more reliable is the goodwill and mutual sense of responsibility that exists among neighbours who are bound together into a community, their good intentions enforced by social pressure and the intimacy of long association.

The meal process are complete to determine toe lower cost health processes this depend on the farmers economy because he is provide the meal to the market when according to market need but not meat good relivilities of the former. This is big reasons of our economy.

Compassion usually only extends as far as the eye can see. We live, most of us, oblivious to such things as world hunger, deforestation, and toxic waste – oblivious, that is, until it comes a'knocking. Our response to a starving person at the door is different from our response to a starving person in Africa, who we know exists, in theory, but who isn't in our faces. How could we expect consumers, then, to really care about farmers, who are separated by layer after layer of distribution, And how could we expect farmers to really care about the wholesomeness and goodness of their meal, when the beneficiaries are similarly remote, and when the tangible rewards hinge not on goodness but on cost efficiency? The process is completed than automatically packaging possible of that processes.

Social connections, and human contact, are the ally of good intentions. It is easy to participate in an exploitative food system when you cannot see the victims. But it goes against human nature, and rational self-interest, to victimise someone with whom you have a continuing relationship of mutual dependency. In the arena of meal, we put it this way: "Social connections and human contact are the ally of good meal."

An alternative path exists: meal should not be primarily a commodity. Meal is a gift of God's Good Earth, for which all religious traditions teach gratitude. To subject it to the economic regime of the lowest bidder is to desecrate the gift and insult the Giver. For most of human history, the sharing of meal was a significant social act, cementing ties between friends and kin, showing welcome to strangers. Today it has become an anonymous act of commerce.

Other people in other times would no doubt have thought it exceedingly strange, if not downright obscene, for total strangers to grow, process, and even cook nearly all one's meal.

The Proper Role of Money

That is not to say that meal should never be bought. Money has its rightful role, even among friends, as an aid to fairness and a means of support. What I am saying, rather, is that the sharing of meal should be part of a personal relationship. Money may be involved, but the income motive should be secondary. In my economic relationships with the local farmers I know, I am happy to pay them a fair price, in hopes that they will be prosperous. My sentiment is partly selfish, because I know that if they are prosperous, they will continue to provide me with good meal. But also I simply don't feel good about eating food that came through the devaluing of another human being's labour, especially when I know that human being personally. When a personal relationship exists between meal supplier and meal consumer, then bargaining becomes a process of each party coming to understand the other's circumstances to find a mutually fair price, rather than a heartless and shameless exercise of getting the best possible price, which in economics is called "maximizing utility" and in common sense language is called greed.

In production as well, true organic farming that builds the soil, doesn't depend on large external inputs, doesn't pollute the environment, and doesn't exploit labour is again incompatible with higher level production of crops or livestock. The compromises farmers make to achieve the economies of scale necessary for survival in the commodity economy are, again, not entirely gratuitous. No one builds a 5000-hog concrete hog barn out of aesthetic delight, devotion to animal husbandry, or desire to nourish one's neighbours with good meal. The connection between fresh, wholesome meal and localism is not an incidental one.

The people can buy the soda but not know "How to make it". Although we will teach and determine the various factor when it is use to make for the quantities. With more efficiency with maintain the quality.

We envision a society also where farmers are personally acquainted with the people who eat their produce, or perhaps, for certain products, linked through one degree of separation. This is workable, because almost as if by design, the ideal size for a sustainably operated mixed family farm is sufficient to meet the food needs of 20 or 30 families. Of course, farms might specialize to some degree, so each family might patronise three or four farms; even so, this calculates out to a manageable number of people per farm, few enough that the farmer can know each personally. Personally I believe that true sustainability requires even smaller farms, and more farmers. Maybe almost everyone not living in a city should be a part-time farmer, at least to the extent of tending a vegetable garden or keeping a few chickens.

AUTHENTIC CHANGE IS MORE THAN A BRAND SWITCH

It is essential to realise that positive dietary change involves more than choosing product A over product B from the supermarket shelves, or supplement pill A over supplement pill B. If health may be defined as wholeness, then it is to be expected that a move towards healthier meal will involve a move towards wholeness in other areas of life as well. Healthy eating is not so simple as a brand switch, because mass distribution and mass marketing generally rests on the same foundation of ill health, where money has exceeded its proper facilitative role to become an end in itself, supplanting other forms of human relationship. The pathological state of food economics inevitably expresses itself in the meal. Checking ingredient lists in packaged meals is important, but it only goes so far. Seek instead to buy meal from real people that you feel good about.

Because the social function has entered the real of money and not other excuse that we are getting the price this is not run ling time, as food and other material. Writing about the Ladakh of India, Helena Norbert-Hodge observes that before the road was built connecting them to the outside, money was used but rarely, for such things as jewellery. The necessities of life were shared or exchanged through other social mechanisms. Money was not important, because it bought nothing important. Today] the opposite is true: everything costs money. Even such intimate functions as housecleaning, cooking, and child care are given over to paid outsiders. In today's society, unlike ever before, without money one has nothing. One is completely dependent on money to live. Or so it appears. No wonder, then, the omnipresent dread of my students at Penn State that they will be unable to find a job and "make a living." When I poll them about their main motivation to attend college—to get a good-paying job, or to learn things – about 90 per cent acknowledge the former. Their mayor life decisions are determined by money. Their thralldom to it is nearly complete. And they are not to be blamed for this, in a world where money has usurped most other modes of economic relationship.

I must emphasize again that I don't advocate abandoning money altogether, just keeping it in its proper place. Specialization is fine when it comes to the technical products of industrialised society, but in the intimate realms of life, specialisation has run amok and stolen something of our humanity. Meal is one such realm. Here we need less specialisation and more sharing, more producers and fewer middlemen. An anonymous, commodity-based meal system is inconsistent with fresh, living, wholesome meal. Of necessity it impoverishes the farmer, sickens the consumer, and ruins the land. As a consumer, to begin withdrawing from this system will mean paying more, initially, for certain mealstuffs, and spending more time in the kitchen. The good result, though, will be to enrich you. By withdrawing, partially, from the money economy in this one area of life; that is, by basing economic decisions on something other than the best price, you will quite naturally become richer in those things that money cannot buy. And, almost as if by magic, you will become materially more wealthy too—not because of your bank balance, but because you will feel more secure in the material world.

Continuous Fermentation Process

Continuous chemical change by fermentation are those in which fresh nutrient medium this is added either continuously or intermittently to the chemical change by fermentation vessel, accompanied by corresponding continuous of portion of the medium for recovery of cells or chemical change by fermentation products.

A continuous chemical change by fermentation can be conducted in various ways. It can be carried out as a single stage in which a single fermenter is inoculated then kept in continuous operation by balancing the input and output of nutrient solution and harvested culture, respectively. In a recycle continuous chemical change by fermentation a portion of the withdrawn culture, or of the residual unused substrate plus the withdrawn culture, is recycled to the fermentation ban be recycled for further microbial attack. A portion of the organisms being produced during a continuous chemical change by fermentation also can be recycled in certain instances in which the actual available substance is sulfite waste liquor with its low available carbohydrate content in this instance the recycling of cells provides a higher population of cells in the fermenter and hence, a greater productivity. Multistage ethanol production with recycling of forth appearing in fermentation is another such example.

A third type of continuous chemical change by fermentation the multiple stage continuous chemical change by fermentation. It involves two or more stages with fermenter being operated in sequence. The multiple stage continuous chemical change by fermentation is particularly applicable to those chemical change by fermentation in which the growth and synthetic activities of the cells are not simultaneous that is synthesis is not growth related but occurs after the cell-multiplication rate has slowed.

Continuous chemical change by Fermenting Process

Growth of micro-organisms during batch chemical change by fermenting confirms to the characteristic growth curve, with a lag phase followed by a logarithmic phase. This, in turn, is terminated by progressive decrements I in the rate of growth until the stationary phase is reached. This is because of limitation of one or more of the essential nutrients.

In continuous chemical change by Fermenting, the substrate is included to the fermenter contiguously at a fixed rate. This maintains the organisms in the logarithmic growth phase. The fermentation products are taken out continuously. The design and arrangements for continuous chemical change by Fermenting, are some what complex.

Fermented products from fish

Fermentation involves the breakdown of proteins in the raw fish to simpler substances which are themselves stable at normal temperatures of storage. In some processes where it is controlled by adding salt, only a partial breakdown of the protein takes place so that a desired type of flavour is produced simultaneously ensuring preservation of the product. Breakdown of proteins is brought about by the action of enzymes and sometimes microorganisms are also involved.

Fermentation technology for the preservation of fish has been very popular from time immemorial. Fermented fish pastes and sauces are relished as a condiment (flavoured salt) along with cooked rice in many of the South East Asian countries. On daily basis, nobody consumes large quantity of these products but almost every one consumes a little bit every day.

Types of fermented fish products

Products in which the fish retain substantially their original form or preserved as large chunks. Examples: pedah siam (Thailand), makassar (Indonesia), buro (Philippines), Colombo cured mackerel (India)

Products in which fish are reduced to a paste. Examples: ngapi (Burma), prahoc (Kampuchea), belachan(Malaysia), trassi (Indonesia), bagoong (Philippines)

Products in which fish are reduced to a liquid. Examples: budu (Malaysia), patis (Philippines), nuoc-mam (Vietnam), nampla (Thailand)

Traditional fermented fish products are basically salt fermented products. Depending on the proportion of salt added, the products can also be classified into high salt (more than 20% of total weight), low salt (6 to 8%) and no salt products

The dominant flavour giving components of fermented fish products are proteins and their hydrolytic cleavage products such as peptides, peptones, amino-acids; higher fatty acids and their esters; glycerides and their derivatives, monosodium glutamate, nucleotides and inosine monophosphate

Nutritive value of fermented fish products

Fermented fish sauces and pastes generally contain amino acids and polypeptides equivalent to about 10% protein. Amino acids occur in such products without much change in composition and quality with respect to fish and hence contribute towards nutrition just like fish protein. They are good sources of calcium, iron and some B group vitamins. However, nutritional importance of these products is limited by its high salt content which restricts its bulk consumption. Moreover, these traditional products are used as a condiment rather than to derive nutrition.

Fermentation involves the breakdown of proteins in the raw fish to simpler substances which are themselves stable at normal temperatures of storage. In some processes where it is controlled by adding salt, only a partial breakdown of the protein takes place so that a desired type of flavour is produced simultaneously ensuring preservation of the product. Breakdown of proteins is brought about by the action of enzymes and sometimes microorganisms are also involved.

Fermentation technology for the preservation of fish has been very popular from time immemorial. Fermented fish pastes and sauces are relished as a condiment (flavoured salt) along with cooked rice in many of the South East Asian countries. On daily basis, nobody consumes large quantity of these products but almost every one consumes a little bit every day.

Byproducts

Fish processing and filleting industries turn out large quantities of fishery waste which are good sources of high quality protein, fat minerals etc. The shellfish processing industry is also a major source of waste comprising of head and shell. The proper disposition of such fish and fishery waste is achieved by processing these wastes into different products intended for human consumption, animal nutrition or industrially useful products. These are generally termed as byproducts.

Fishmeal

Fishmeal is a highly concentrated nutritious feed supplement consisting of high quality protein, minerals, vitamins of B group and other vitamins and other unknown growth factors. It is produced by cooking, pressing, drying and grinding the skeletal remains along with the adhering proteinaceous tissues of fish from filleting or canning operations or by processing whole miscellaneous fish mainly caught along with prawns, which include Jew fish, Sole, Silver bellies, Ribbon fish and the like.

The composition of fishmeal differs considerably due to the variations in the raw materials used and in the processing methods and conditions employed. Fishmeal is rich in all essential amino acids, B group vitamins and minerals particularly phosphorous and calcium and therefore finds an important place as a feed supplement for poultry and cattle.

The main raw material for fishmeal is the abundant but sporadic catches of oil sardine on the West Coast. In addition to this the fishmeal plants also draw some of the miscellaneous fishes like silver bellies, flat fish, ribbon fish, juveniles of some commercially important fishes, wastes from fish canneries etc. as raw materials for fishmeal and oil production.

Traditional fishmeal production in India was from the sun-dried fish collected from various drying centers all along the coast and the product was chiefly used as manure. Better quality fishmeal has been a prominent item of export from the very beginning of this industry. The importance of improving the quality for better use was felt and the Ministry of Food and Agriculture has, as early as in 1959, laid specifications regarding the quality of fishmeal. Later the Indian Standards Institution has brought out the Indian Standards Specification (IS: 4307-1967) for fishmeal as livestock feed for facilitating proper quality control.

Manufacturing process

Fish can be reduced to fishmeal by two general processes:

Dry-rendering - Dry-rendering or dry-reduction process is suitable only for lean or non-oil fish such as silver bellies, Jew fish, sciaenids, ribbon fish, sole, anchoviella, carcasses of shark etc., fish offal and filleting waste. In this process if the quantity of fish processed is very small, it is dried to a moisture content of 10% and pulverised. If the quantity to be handled is sufficiently large a steam-jacketed cooker-dryer equipped with power driven stirring device is used. Sometimes, if the size of fish is comparatively large a coarse grinding is also done before being fed into the cooker-drier. The cooker drier may be operated at atmospheric pressure or under partial vacuum.

Being a batch operation the process will have only limited capacity and labour costs will be high. However, the water-soluble materials are retained in the meal. It has the additional advantage for changing the operating conditions like cooking time and temperature depending on the raw material handled.

Wet-rendering - Wet-rendering or wet reduction process is normally applied to fatty fish or offal where simultaneous production of fishmeal and fish body oil is envisaged. The process consists of grinding, cooking to soften the flesh and bones and to release the oil, pressing to expel the liquor and oil, fluffing the press cake, drying, grinding and packing the meal, centrifuging the press liquor to remove suspended particles and to separate oil and concentrating the stick water. The process requires elaborated equipment and is normally a continuous one and therefore adaptable to the reduction of large quantities of fish.

In a continuous wet reduction process the coarsely ground or fresh raw fish or offal is passed through a stationery horizontal cylindrical cooker by means of a screw conveyor at a predetermined rate. Steam is admitted into the cooker through a series of jets. While the cooked mass is pushed forward by the conveyor the expressed liquid is allowed to drain through a screen on the bottom half of the cylinder. The cooked mass is passed through a continuous screw press. The press cake is fluffed and dried to a moisture content about 8%. The suspended fishmeal present in the press liquor is separated by centrifugal sedimentation and the oil by centrifugation or other conventional methods.

Fish liver oil

From time immemorial fish liver oils were used for the treatment of vitamin A & D deficiency diseases. The therapeutic values of fish liver oil was discovered in the 18th century and fish liver oil became a common medicinal product. Both vitamin A and D are found in certain fish liver oils. The most important fish liver oils are obtained from cold, haddock and shark. Halibut and Tuna livers are also rich sources of vitamin A & D. The weight of liver, fat content and presence of vitamins are dependent on a number of factors like species, age, sex nutritional status stage of spawning, and area from where it is caught.

In Cod (Gadus collarius), Coal fish (Pollahius vireus), and Haddock (Melanogramms aenglefinus), the weight of liver normally amount to 4 to 9% of the whole fish and livers contain about 45% to 67% oil. The species of shark such as dog fish (Squalus acanthias), Greenland shark (Somniosus microcephalus) and basking shark (Certorhinu maximus) have large fatty livers weighing up to 10 – 25% of the whole fish containing 60 – 75% oil. But halibut, tuna and whale have 1 % liver having 4 to 25% oil with high vitamin A & D potency.

Elasmobranch fishes, Sharks, Skates and Rays constitute a good percentage of the total marine landings in India. Of these, sharks possess commercial importance owing to their liver oil content. Sharks are available in the east and west coasts of India and their size vary from 25 cm to about 600 cm. Of the 37 species of sharks available in Indian waters, only certain varieties, viz. Tiger shark, black fin shark, hammer headed shark and saw fish, are commercially important, and yield liver oil with vitamin A content. The potency of vitamin A has been reported from 1000 to 3,53000 International Units per gram and oil content up to 80% weight of liver. With decreasing fat content, the colour of liver changes from yellow to brown and texture from soft to firm. Shark liver oil is valued for highest unsaponifiable matter in the form of squalene which is present in the liver of certain species of deep sea shark. Apart from liver oils, shark fins and teeth are also items of commercial importance.

Depending on the oil content and vitamin A potency fish livers are generally classified into three groups:

- Low oil content – High vitamin A potency
- High oil content – Low vitamin A potency
- High oil content – High vitamin A potency

Preservation of livers for oil extraction

The fish is to be eviscerated and livers processed as quickly as possible, especially in warm climates to obtain high quality, light coloured oil with good flavour and odour having minimum free fatty acids (FFA). The FFA content of good quality liver oil should be less than 1%. The spoilage of livers may be

- Chemical spoilage causing rancidity
- Biochemical spoilage caused by enzymes
- Microbial spoilage resulting in fermentation or putrefaction.
- Before processing, the livers should be freed from gall bladders and other visceral matter and washed to remove slime.

Processing

The processing procedures of fish livers without affecting the quality of the oil extracted can be summarised as:

- Steaming
- Solvent extraction and
- Alkali/ednzyme acid digestion.

The process selected should depend on the vitamin and oil content of the livers. Depending on the oil content and vitamin A potency, fish livers are generally classified into three groups as :

High Oil – Low Vitamin A Livers : In such livers the large volume of oil present acts as a solvent for vitamin A thereby facilitating the extraction. The method of heating the fish livers with steam, either directly or indirectly was employed for extraction. Cooking results in the thermal rupture of the cells and releases the oil. If heated indirectly in steam jacketed kettles, it should be stirred mechanically to facilitate digestion.

Low Oil –High Vitamin A Livers : For such livers which maintain only very low oil but high potency vitamin the direct or indirect steaming procedure will not release the oil without degradation of vitamin A. In such cases the protein is digested and solubilized to release the enclosed oil and vitamin A or a solvent is used to extract the oil. The different methods of digestion employed are:

- Alkali digestion
- Enzyme and alkali digestion
- Acid digestion.

Mild alkali digestion is the easiest and most general practice followed in many establishments. The process involves addition of 1 – 2% by weight sodium hydroxide or 2 to 5% sodium carbonate to the disintegrated livers and heating to 80-90° C with live steam while stirring. This results in the release of vitamin A and oil from the cell. The digested liquor is centrifuged while hot, to separate the oil.

Livers with very low oil content which do not yield their oil directly are minced and covered with any suitable oil and heated, with stirring, to extract the oil soluble vitamins and centrifuged. The process is repeated for further recovery of oil and vitamins.

Heating the livers with steam applied either directly or indirectly has been generally employed for the large scale extraction of oil. For this purpose various processes/equipment are used in different countries. For small scale and cottage scale extraction of good quality liver oil, the following method may be adopted:-

The livers are chopped into small pieces. They are then mixed with ½ to 1 part of water and heated in a cooking vessel described in the extraction of fish body oil to a temperature of 80-90°C for one hour. The mixture is stirred well and heating controlled to keep the mass around 90°C. The floating oil is transferred to a separating vessel. The residue is taken in a canvas bag and pressed in a screw press and the press liquor is collected for separation of the residual oil.

The oil water separator is a galvanized iron cylindrical vessel with two outlets, one at the bottom, and one at half way up the side fitted with stoppers or valves. The oil water mixture collected during cooking or from the screw press is taken in the separator and allowed to stand for about 30 minutes when the oil and water separates in two layers, the oil remaining at the top. After separation, the oil is run off through the side outlet, a suitable oil level being maintained by adjustment of the water level in the tank by means of the bottom outlet. The equipment is simple and easy to construct and can be built by any sheet metal workshop.

High oil content with high hydrocarbon content (chiefly squalene) : Squalene a highly unsaturated aliphatic hydrocarbon, is present in certain shark liver oils, mainly of the family Squalidae, cod and some vegetable oils like olive oil, wheat germ oil, rice bran oil. Chemically it is known as 2, 6, 10, 15, 19, 23, hexamethyl-2,6, 10, 14, 18, 22 tetracosahexaene having a molecular weight of 410.70. it is an isoprenoid compound containing six isoprene units.

2

Fermentation Technologies and Types of Associated Fish Products

Throughout history, different cultures have adopted a range of techniques to slow down the bacterial and enzymatic degradation of fish including salting, drying, smoking, and fermentation. Fermentation is one of the most ancient and significant traditional methods of food preservation. It entails the production of alcohol, lactic acid, and carbon dioxide from food sugars, primarily through the action of microorganisms and enzymes, resulting in foods with distinct quality characteristics such as aroma, taste, shelf life, texture, and nutritional value that differ significantly from the raw material. Fish sauces and fish pastes are the most popular fermented fish products in many Asian countries and are regarded as indispensable condiments in these culinary cultures. Fish sauce is produced due to changes in physical and chemical characteristics brought about by microbial activity in high salt concentrations and low oxygen levels. Another form of fermented fish product is one where the fish retains its solid form and does not get reduced to a paste or a liquid form. Based on the country and its traditional practices, fish sauce is made by combining fish and salt in a 2:1 to 6:1 ratio.

Traditional and non-traditional fish products are classified in different ways according to their respective mechanism of processing, substrates used, salt concentration, and the form of the final product. Saisithi has divided such processes and their resultant fermented fish products into three categories. The first is a purely traditional and natural process based on enzymes present in the fish flesh and entrails with salt as seasoning. In the second category, microbial enzymes and salt are added in the form of 'starter cultures' on the flesh of fish and entrails. The third category consists of non-traditional products manufactured by artificially induced accelerated fermentation. Rattagool classified traditional fermented fish items in Southeast Asia based on the substrates used or added during the fermentation processes, such as salt and carbohydrate. Saisithi proposed a classification system for traditional fermented fish products based on the substrate and source of the enzymes used in the fermentation that comprises three groups: (1) where carbohydrates are not a part of the fermentation and the product is the result of the activity of the bacterial enzyme already present in the fish and salt mixture; (2) where carbohydrates are present in the fish and carbohydrate mixture used in the fermentation process; and (3) where a 'starter culture' is used that brings about the fermentation of the carbohydrates. The most usual process, however, is the first one where the fermentation is brought about by the enzymes present in fish.

Ruddle and Ishige gave a well-defined generic classification and methods of preparation of fermented fish of East Asia. They classified fermented fish into two groups. In one, freshwater fish was used for fermentation, and the other consisted of marine finfish and shellfish. Normally, traditional fermented fish products are salted products with different salt concentrations. Accordingly, fermented fish can also be classified into highly salted products with 20% salt, low salt products with 6–8% salt, and products without any added salt. Fermented fish products are also categorized according to the form retained by the final product. The main types of fermented fish according to the form are the following: (a) Fermented fish products retaining their original form which are often salted, like buro (Philippines), makassar (Indonesia), pedah-kemburg (Thailand), makassar (Indonesia), ngari(India), Colombo cured mackerel (India); (b) products that contain fish that have been reduced to a liquid, e.g., patis (Philippines), budu (Malaysia) nampla (Thailand), and nuoc-mam (Vietnam); and (c) Products in paste form, e.g., ngapi (Myanmar), bagoong (Philippines), prahok (Cambodia), trasi (Indonesia), and belacan (Malaysia).

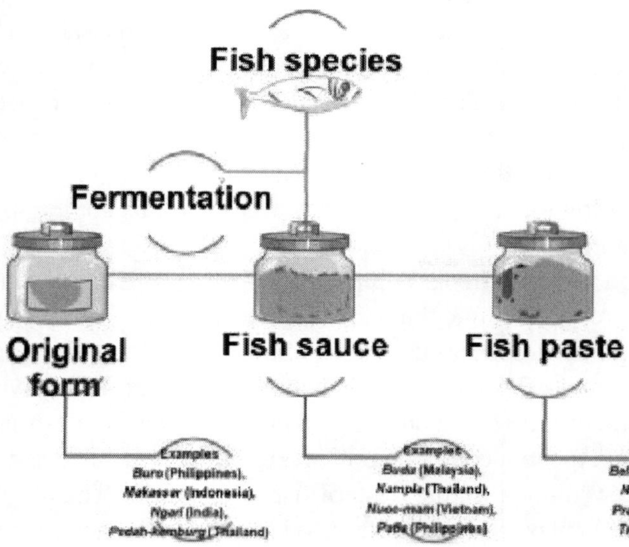

Fig: Fermented fish products in various forms that are considered a delicacy in Asian cuisine (Rao ; Beddows)

Fermented fish is significant traditionally and economically in East and Southeast Asian countries. Because of its unique flavor and taste, its popularity in these countries is far reaching. It is a common food in northern Laos, Cambodia, southern Myanmar, northeastern Thailand, and the region of Luzon in the Philippines. All three forms of fermented fish products are prevalent in these countries. Again, in many fermented fish products of Asia, koji (Aspergillus sp.) is added to enhance the fermentation. The Asian climate has a higher humidity and temperature that is optimal for filamentous fungi like Aspergillus, Penicillium, Rhizopus, Mucor to thrive, leading to their wide use in fermentation processes in various parts of the continent

In Thailand, well-known fermented fish products are budu, kapi, and prala. Budu is an intermediate between fish sauce and fermented fish paste and has a colloidal consistency, and prala is fermented fish paste where rice bran is added as a carbohydrate source. Kapi, on the other hand, is a fermented shrimp paste using salt in a ratio of 5 (shrimp):1 (salt), that is then dried and compacted. The technology used for the preparation of these products may be similar, but the additives are different depending on the product, with ingredients such as rice and fruits being used in many places. The use of an additive facilitates the growth of lactic acid bacteria, lowering the pH of the solution. Sometimes, similar products may also have diverse names in different localities of Thailand.

The traditional fermented fish products of Malaysia are: pakesan prepared from freshwater fish mixed with tamarind and salt, belacan (shrimp paste), cincaluk, budu, and kikapikan. In the Philippines, fermented fish products produced with lactic acid bacteria are consumed in the Central Luzon area. Bagoong (fish paste), balao-balao (fermented shrimps), burongisda (fermented rice and fish), bagoong alamang (shrimp paste), and paite (fish sauce) are some of the most common and popular fermented fish products in the Philippines. It was reported that many fishes, including herring and anchovy, and shellfish, such as shrimp, are used for the preparation of bagoong by mixing salt with well-drained samples. Bagoong has a characteristic salty taste, faintly cheese-like odor, and prepared from fish or shrimp. The fish sauce is called paite and obtained from thoroughly fermented bagoong. Balao-balao is a dish made with cooked rice and shrimp (Penaeus indicus or Macrobrachium sp.), while burongisda is made with freshwater fish and rice.

The most popular fermented fish products in Indonesia are pedah (fatty, partly dried, salty fish), trasi (shrimp paste), bekasang, and bekasam (freshwater fish that has been fermented), jambal roti (marine fish that has been fermented and then dried and salted), budu, cincaluk, naniura, petis, picungan, pudu, rusip, and tukai. Fish products that have undergone fermentation (terasi, peda, and kecapikan) account for less than 2% of all fermented fish products processed.

Myanmar's traditional fermented fish ngapi made from salted fish or shrimp is considered the cornerstone of Burmese traditional food. Nuocmam is a fish sauce consumed by a huge majority in Vietnam. It is prepared from small fish and shrimps. The fish are layered with salt and kept for fermentation. The supernatant liquor is decanted after it has aged to produce high-quality nuocmam. During the fermentation process of fish sauce, the fish tissue is extensively liquefied, and the resulting fluid portion is collected as the sauce.

Fermented fish sauce, like yu-lu which is commonly consumed in China as a condiment, is a popular and indispensable seasoning in Chinese cuisine. Anchovies (or other small marine fish) are allowed to ferment by adding salt at the level of 30–40%, salt/fish (w/w) in this product. The period of fermentation is 12–18 months. In Japan, the fermented fish is called shiokara (preserves the whole or partial of the fish shape), and shiokara paste (yielded by pounding or grating). When, along with fish, additives like salt, vegetables, or steamed rice are used, then it is known as narezushi. Shiokara and shiokara pastes are mostly used as condiments and side dishes.

Korea also possesses a rich dietary tradition of fermented food and fish. Some of the major fermented fish products of Korea are jeotkal, eo-ganjang, and sikhae. In the context of India, the practice of pit-curing is widespread, particularly along the country's coastlines. The salted fish are buried in pits lined with mats for a period spanning from a few days to two weeks or more in this curing method. The finished product is partially dry and has a distinct flavor and taste that is popular in the eastern region of Tamil Nadu, India.

The 'Colombo cured' method is a prevalent practice in Southern India and Sri Lanka for the preparation of fermented fish. Typically, this method involves mixing mackerel fish and salt in a 3:1 ratio in a concrete tank, along with dried tamarind. Improvements in the quality are obtained if 5% acetic acid is added in place of tamarind.

The fish curing industry has a large presence in India, but paradoxically, it is also one that lacks mainstream commercial visibility though it is popular as a local cottage industry in the northeastern and coastal parts of India. Fermented fish products, again, are quite prevalent in North East India and West Bengal. It is interesting to note that many such products, like shidal and ngari, do not use salt. Salt, which has a long and contentious colonial history in India, was introduced in the Northeastern region by the British people. In place of salt, sometimes a local alkali called khar is used, which helps in alkaline fermentation (napham, nichow and nakham). The fermented fish pastes prepared in this region are pounded with various herbaceous plants like taro and allowed to age inside bamboo stems. Bamboo is abundant in this region, which has the highest diversity of bamboo species in India. The use of bamboo stems and leaves in cooking is a widespread practice in Northeast Indian communities that finds its way even into aspects of folk cultures and studies of the region. Cooking food or processing it in bamboo stems imparts a unique flavor to the food. Thus, the fermented foods found in this region are unique and one-of-a-kind to a great extent due to the use of bamboo in their processing. Two types of fermented fish products have been reported in North East India: one that retains its solid form after fermentation and the other in a paste form. They are used as condiments or with other vegetables for their texture and flavor. The local people also associate these products with some health benefits. North East India's fermented fish products are diverse.

Shidal is a popular fermented fish product found in different states of Northeast India. It is prepared from Puntius sp. (generally Puntius sophore) and Stipinna phasa and is known by different names across five states of Northeast India, namely Assam, Mizoram, Arunachal Pradesh, Nagaland and Tripura. Ngari is a traditional Manipuri dry, unsalted fermented fish product made solely from sun-dried Puntius sophore Ham. Tungtap is traditional fermented fish in Meghalaya which is prepared from Puntius sp. and Danio sp.. Hentak, numsing, sukakomacha, sukati, napham, nichow, nakham are some other fermented fish products which are prepared by different ethnic communities found in Northeast India. Lonailish is salted and prepared exclusively from Indian shad (Tenualosa ilisha) popular in both Bangladesh and parts of India bordering Bangladesh.

MICROBIOTA OF FERMENTED FISH PRODUCTS

Microorganisms are an integral part of all fermented food products. Lactobacillus and other microorganisms in fermented food have gained the status of GRAS (Generally Recognized As Safe) which has enhanced the commercial importance of fermented food products due to the beneficial physicochemical properties endowed by these microbes. A broad spectrum of microorganisms has been reported from different fermented fishery products of Asia. The diversity of microbial strains is quite interesting, and their presence is not uniform. In the present study, Pseudomonas aeruginosa and Escherichia coli have been reported in only one product (Pedah). Similarly, Staphylococcus aureus was discovered in five different products, including pedah, belacan, ngari, tungtap, and shidal. However, we discovered that the most prevalent genera are Lactobacillus, Staphylococcus, and Bacillus, which are found in 12 of the 18 products included in this study. Micrococcus and Staphylococcus were found in 9 products, making them the second most prevalent bacterial genera in fermented fish. Furthermore, we discovered that Pseudomonas, Flavobacterium, and Brevibacterium were less dominant strains, with their presence recorded in five, three, and two products, respectively.

Suanyu is a fermented fish product popular among the indigenous people of Southwest China. Suanyu's dominant bacteria were Lactobacillus, Tetragenococcus, and Weissella, contributing 53.99%t, 35.60%, and 4.10%, respectively. In budu, the strains of bacteria isolated were Tetragenococcus halophilus, Bacillus licheniformis, Staphylococcus piscifermentans, and Bacillus subtilis. In pedah, Lactobacillus sp. and Staphylococcus sp. were reported (Table 1). In ngapi, Bacillus sp. Staphylococcus sp., Staphylococcus aureus, Micrococcus sp., and Pseudomonas sp. were reported. In nuoc mam, a fermented fish from Vietnam, microbial species like Staphylococcus sp., Bacillus sp., Pseudomonas sp., Pediococcus sp., Lactobacillus sp., Streptococcus sp., Halobacterium cutirubrum, Halobacterium salinarum was isolated.

These studies suggest that Bacillus sp., Staphylococcus sp., Lactobacillus sp., Micrococcus sp. have a specific role in fish fermentation. In all these fermented fish products mentioned above, the raw materials used and the manufacturing processes differ, and these factors influence the type and number of microorganisms present in them. In some cases, the addition of certain microorganisms as 'starter cultures' leads to the acceleration in fermentation time resulting in a better product.

The addition of proteolytic enzymes of Aspergillus oryzae and Aspergillus niger hastened the production of bagoong and patis, respectively. Staphylococcus sp.SK1-1–5 and Virgibacillus sp. SK 33 strains, when used as 'starter cultures,' increase the desired volatile compounds in the fish sauce by inducing a higher production of proteinases. It has been claimed that Tetragenococcus halophilus improves the flavor of fish sauce by producing volatile compounds like benzaldehyde, 2 &3- methylbutanal, and 2-methylpropanol that give the desired flavor to the fish sauce. Fungi like Aspergillus luchuensis have been attributed with generating a superior flavor due to their proteolytic and lipolytic activities in low salted fermented fish in a study by Kim. Thus, endogenous enzymes found in a wide spectrum of microbes contained in fermented fish products can impart desirable organoleptic properties and prolong their shelf life. The balanced metabolites present in fermented fish can determine the preferable flavor and texture. Furthermore, the microorganisms also degrade the anti-nutritional components and produce metabolites with less toxic in nature. Histamine, for example, is a toxic amine found in a variety of fermented fish products that has serious health consequences. In a study by Kuda , Tetragenococcus halophilus isolated from nukazu had a suppressive effect on histamine accumulated in salted fermented fish. Weissella strains from plaa-som produced antibacterial compounds and folate. In the analysis and review of 18 different fermented fish products of South and Southeast Asia, it was found that Bacillus, Lactobacillus, Micrococcus and Staphylococcus were the most predominant.

Biochemical characterization of fermented fish in Asia

In the process of fish fermentation, biotransformation of proteins and fatty acids takes place as these are broken down into simpler compounds by microbes or the enzymes produced by them. The fish also contain indigenous enzymes that can bring about the breakdown of fish protein that cause the textural change in the final product. South Asian fermented fish products are alkaline. In alkaline fish products, the hydrolysis of proteins into peptides and amino acids takes place naturally, releasing ammonia. The release of ammonia raises the final product's pH. The alkaline pH and ammonia in these products control the growth of a few dominant bacteria, which allows the anaerobic breakdown of proteins that release amine compounds. The high pH, free ammonia, and the rapid growth of essential microorganisms do not allow the sustenance of those bacteria that may spoil the product. Thus, the product is very well preserved and stable when dried. The majority of alkaline fermentations occur spontaneously in mixed bacterial cultures, with Bacillus subtilis predominating. The physico-chemical characterization of various types of fermented fish is dependent on a variety of factors, including the type of processing fish used, the percentage of salt in the product, the number of days of fermentation, and the presence of microorganisms in the product.

The presence of amino acids, both essential and non-essential, in fermented fish products has been widely reported. Glutamic acid and aspartic acid were the most abundant amino acids in rusip, an Indonesian fermented fish. Histidine, serine, glycine, arginine, threonine, tyrosine, methionine, alanine, and valine were among the other amino acids found in rusip. Glutamic acid contributed to the production of the characteristic flavor and aroma in this fermented fish product. In hatahata-zushi, a Japanese fermented fish, the fatty acid profile consisted of elaidic acid (C18:1n-9), palmitic acid (C16:0), docosahexaenoic acid (C22:6n-3), eicosapentaenoic acid (C20:5n-3), and PalC16:1n-7. The dominant fatty acids reported in ngari were oleic acid (23.58%), palmitic acid (C16:0), and vaccenic acid (C18:1n-7). Fatty acids reported in Hentak were stearic acid (C18:0), palmitoleic acid (C16:1n-5), and oleic acid. There were 62 volatile compounds identified and reported in Chinese traditional fermented shrimp, including, aldehydes, alcohols,, acids, ethers, esters, pyrazines, hydrocarbons, ketones, and phenols. The literature suggests that the production of flavor and texture of fermentation fish depend on the amino acids and fatty acids that have been formed after microbial decomposition.

HEALTH AND SAFETY CONCERNS OF FERMENTED FISH

Fermentation of fish imparts desirable organoleptic characteristics and improves mineral bioavailability in the product. Fermented fish contains significantly lesser amounts of fatty acid compared to fresh fish. Fermented fish oil has been found to contain particularly high levels of EPA (eicosapentaenoic acid) and DHA (docosahexaenoic acid). Han et al. reported that fermented fish oil contains DHA which can help alleviate the symptoms of atopic dermatitis. Fermented fish also contains natural antioxidants and essential nutrients. It has been reported that bioactive peptides derived from fermented fish products act as antioxidants. According to Siddegowda et al. , bioactive peptides may exert a variety of biological functions depending on their amino acid sequences, including antioxidant, antagonist, antihypertension, immunomodulatory, antithrombotic, antimicrobial, and anticancer activity. Aoshima and Ooshima investigated the antioxidant activity of two Japanese liquid condiments, shoyu (soy sauce) and gyoshoyu (fish sauce). Siddegowda et al. stated that the Labeo rohita head sauce produced by enzymatic and fermentation methods possesses in vitro antioxidant properties Antioxidants are critical for human health, as reactive oxygen species are thought to be a major cause of aging and carcinogenesis. Siddegowda et al. examined the sauce extracted from Rohu (Labeo rohita) enzymatically and fermentatively. They discovered that the total antioxidant activity (as ascorbic acid, g/ml), 2,2'-diphenyl-1-picryl-hydrazyl (DPPH, %), and 2,2'- azinobis-3-ethylbenzothizoline-6 sulfonate (ABTS, %) scavenging activities of fermented sauce samples were significantly higher at 3441±?60, 49.57?±?1.7, and 99.0?±?0.3, respectively, when compared with enzyme treated samples which showed 803?±?13, 28.60?±?1.1, and 78.4?±?1.0, for the respective parameters, after 180 days storage at 37°C. Some research has suggested that fermented fish can produce bioactive metabolites (such as lactoferrin and flavonoids) as well as polysaccharides from raw materials. These bioactive peptides are antioxidative , ACE inhibitors , antihypertensive and have anti-proliferative effects on human lymphoma cells.

Biogenic amines have been reported from many fermented fish products and pose a safety concern. These are nitrogenous compounds made up of amine groups formed by decarboxylation. Biogenic compounds are non-toxic when consumed in lower concentrations but can cause serious health implications when consumed in high amounts. Biogenic amines are vasoactive compounds and may cause an increase in blood pressure, and people sensitive to these amines may show symptoms of allergy, headache, respiratory disorder, nausea, and tachycardia.The presence of elevated histamine and tyramine levels in 47 samples of fermented fish prepared from anchovy, squid, clam, and shrimp suggests that biogenic amine levels in fermented fish products should be closely monitored to ensure human health.

Biogenic amines (BAs) are a significant potential hazard in traditional fermented fish (suan yu), and Enterobacteriaceae are involved in their formation. Amine production capabilities of 97 Enterobacteriaceae strains isolated from traditional foods were examined. The most significant physicochemical factor affecting amine production was pH. These bacteria are capable of adapting to the physicochemical conditions present during the product's fermentation process, posing significant safety risks. Pathogens like Clostridium botulinum, Salmonella sp., Listeria sp., and Staphylococcus aureus were also reported from many fermented fish products which may bear major health implications. It is in such a context that the Cambodian government, as mentioned earlier, saw fit to enforce quality control standards on the nation's representative fermented fish product, prahok.

Cultural significance of fermented fish

This review revealed that certain foods have a lasting cultural significance. South and Southeast Asian countries have diverse and rich food cultures centering on, among other things, fermented food, in particular, fermented fish. However, though Asian foodways have found a global platform, fermented fish figures less prominently in the global food map, precisely because its strong flavor and odoriferous quality usually repel Western sensibilities and taste. There is considerable politics in deciding which food cultures are acceptable for global consumption. Taste itself entails a high degree of cultural politics. Steffan Igor Ayora-Diaz argues that the symbolic aspects of food, and whether it is deemed fit to be consumed, depend on several cultural markers as religion, political and communal ideology, ethnicity, gender, and society as well as transnational factors like the UN and its policies on food security and food safety. For example, in the very recent context of the COVID-19 pandemic, China and the Chinese have received a severe backlash (primarily from people in America and Europe) and regarded as being morally reprehensible due to their food cultures, simply because the practice of consuming bats and pangolins exists in some Chinese communities. In the light of such factors, fermented fish, despite its strong local and overseas base of consumption, has remained relatively obscure in literary and cultural studies. For example, in the realm of the literature, though food is a central theme in Asian diasporic literature, fermented fish by itself is not a central presence even within this framework, partly because these novels have a huge Western readership unfamiliar with this food category. Again, while there is, admittedly, a greater awareness of the health benefits of fermented foods, the same has not translated to its greater inclusion in food guides in many countries around the world. This is not because they are harmful, but rather due to the possibility that they are not part of the food cultures in a particular nation (and therefore unfamiliar) or they are too limited in their production and consumption to become a wider trend or tradition. Thus, there is a significant gap between the relatively limited scientific exploration of ethnic food (including fermented fish) and the embeddedness of fermented fish in the literature and culture of the local communities. Even within food studies, the technical aspects for fermented fish production are more highlighted, with only a few (notable among them being Ruddle, Ishige, Grainger, Tamang and Yankowski) actually connecting it to a larger socio-cultural context. This warrants more extensive research on the interconnections between the scientific and the cultural aspects of fermented fish in Asian foodways.

Whereas fermented foods produced in Europe, North America, Australia, and New Zealand usually depend on defined starter cultures, those made in Asia and Africa often rely on spontaneous fermentation. Moreover, in developing countries, fermented foods are not often commercially produced on an industrial scale but rather in localized cottage industries. Although many fermented products rely on autochthonous microbes present in the raw material, for other products, the introduction of starter culture technology has led to greater consistency, safety, and quality. Microorganisms have already been shown to play an important role in making fermented fish more palatable and valuable. Furthermore, the nutrient composition and taste of the fish product also improve as a result of this process. However, many more traditional fermented fish products that have been consumed for centuries are yet to be thoroughly studied and characterized for their composition. Such products may have numerous isolates belonging to different species with undefined roles. Recent biotechnological tools have the potential to improve food fermentation through the development of novel starter cultures, as well as the addition of value through the fortification of nutrients and probiotics. This will heighten the health benefits associated with fermented food, at the same time promoting new businesses built around it. This, in turn, will assist the growth of cottage, small- and medium-sized businesses, which will be of particular import to Asian economies where there is a presence of significant poverty and economic disempowerment.

3

Industrial Fermentation vs Traditional Fermentation

Industrial fermentation is the intentional use of fermentation by microorganisms such as bacteria and fungi to make products useful to humans. Fermented products have applications as food as well as in general industry.

PHARMACEUTICALS AND THE BIOTECHNOLOGY INDUSTRY

There are 5 major groups of commercially important fermentation:

1. Microbial cells or biomass as the product, e.g. single cell protein, bakers yeast, lactobacillus,E. coli, etc.
2. Microbial enzymes: catalase, amylase, protease, pectinase, glucose isomerase, cellulase, hemicellulase, lipase, lactase, streptokinase, etc.
3. Microbial metabolites :
    - Primary metabolites – ethanol, citric acid, glutamic acid, lysine, vitamins, polysaccharides etc.
    - Secondary metabolites: all antibiotic fermentation
4. Recombinant products: insulin, HBV, interferon, GCSF, streptokinase
5. Biotransformations: phenyl acetyl carbinol, steroid biotransformation, etc.

Nutrient Sources for Industrial Fermentation

Growth media are required for industrial fermentation, since any microbe requires water, (oxygen), an energy source, a carbon source, a nitrogen source and micronutrients for growth.

Carbon & energy source + nitrogen source + O2 + other requirements '! Biomass + Product + byproducts + CO2 + H2O + heat

Buffers: Calcium carbonate, phosphates.

Growth factors: Some microorganisms cannot synthesize the required cell components themselves and need to be supplemented, e.g. with thiamine, biotin, calcium pentothenate.

Precursors: Directly incorporated into the desired product: Phenyl ethylamine into Benzyl penicillin, Phenyl acetic acid into Penicillin G.

Inhibitors: To get the specific products: e.g. sodium barbital for rifamycin.

Inducers: The majority of the enzymes used in industrial fermentation are inducible and are synthesized in response of inducers: e.g. starch for amylases, maltose for pollulanase, pectin for pectinase,olive oil and tween are also used at times.

Chelators: Chelators are the chemicals used to avoid the precipitation of metal ions. Chelators like EDTA, citric acid, polyphosphates are used in low concentrations.

SEWAGE DISPOSAL

In the process of sewage disposal, sewage is digested by enzymes secreted by bacteria. Solid organic matters are broken down into harmless, soluble substances and carbon dioxide. Liquids that result are disinfected to remove pathogens before being discharged into rivers or the sea or can be used as liquid fertilisers. Digested solids, known also as sludge, is dried and used as fertilisers. Gaseous by-products such as methane, can be utilised as biogas to fuel generators. One advantage of bacterial digestion is that it reduces the bulk and odour of sewage, thus reducing space needed for dumping, on the other hand, a major disadvantage of bacterial digestion in sewage disposal is that it is a very slow process.

Phases of Microbial Growth

When a particular organism is introduced into a selected growth medium, the medium is inoculated with the particular organism. Growth of the inoculum does not occur immediately, but takes a little while. This is the period of adaptation, called the lag phase. Following the lag phase, the rate of growth of the organism steadily increases, for a certain period—this period is the log or exponential phase. After a certain time of exponential phase, the rate of growth slows down, due to the continuously falling concentrations of nutrients and/or a continuously increasing (accumulating) concentrations of toxic substances. This phase, where the increase of the rate of growth is checked, is the deceleration phase. After the deceleration phase, growth ceases and the culture enters a stationary phase or a steady state. The biomass remains constant, except when certain accumulated chemicals in the culture lyse the cells (chemolysis). Unless other micro-organisms contaminate the culture, the chemical constitution remains unchanged. Mutation of the organism in the culture can also be a source of contamination, called internal contamination.

Traditional Fermentation Technology

Fermentation technology, as mentioned in the literary texts, is more than 3000 year old in India. The fermentation technology employed a variety of processes and was put to a large number of uses. It also laid the foundation of alchemy and chemistry.

The term fermentation is derived from the Latin word *Fermentum* that stands for boiling. Fermentation is the process of digesting certain substances that leads to chemical conversion of organic substances into simpler compounds. Fermentation technology began with sweet substances (vegetable or animal) in different parts of the world. The process of fermentation was probably discovered by observing the changes in the juices of several fruits and other substances that had been kept for a day or more. It appears that fermentation technology started simultaneously with settled agriculture during the Neolithic period.

A sweet substance known as *Soma* juice prepared by the Vedic Aryans is supposed to be the first product of fermentation in India. The *Rgveda* (c.1500 BC) shows that fermentation technology took its first step in connection with the preparation of *Soma* juice in India. There is also another drink, known as *Sura* (wine/beer), prepared by fermentation. These two preparations have also been used in different medicinal preparations, surgical procedures and in many chemical and alchemical operations. It is believed that acetic fermentation was known to India since the early times.

Vedic Period

The two drinks that are quoted in Vedic texts are *Soma* and *Sura* . The *Rgveda* describes *soma* as a godly concoction, while *sura* is described as mainly a human drink. However, the *Yajurveda* deifies *sura* and uplifts it to the status of *soma* .

Evidence of fermentation in the *Rgveda*

Soma is an invigorating plant product generally prepared by squeezing juice from plants. The following statements give an indication of fermentation:

- Admixture of a thick juice of *Soma* with barley powder (*RV* . IX. 68.4). This statement is a clear indication of fermentation with barley.
- "Fifteenth day old highly intoxicating Soma" (*RV* . X.27.2). In this statement fifteenth day probably refers to the broth fermented in the vat for 15 days.
- "Soma being treated is red" (*RV* .IX.82: 1).

All these statements give an indication of fermentation technology involved in the preparation of Soma.

Evidence of fermentation in the *Yajurveda*

- The Sukla Yajurveda (Vaj. XIX. 13-15; 82-83) describes the formation of two stimulating drinks – Sura and Parisrut. Sura was supposed to be prepared from germinated paddy, germinated barley, and parched rice with the help of ferment. Yeast was used most often as the fermenting agent.
- The Katyayana Srauta sutra (XV, 9.28-30; XIX, 1-2) also gives a complete description of the preparation ofSura . According to this method either boiled rice or boiled barley was mixed with the ferment and also withmasara and the

entire mixture was kept in a jar. The jar was then kept in a pit for three nights. Meanwhile, cow's milk and powdered parched rice were poured in the pit.

Coagulation of Milk

A large number of dairy products that used fermentation technology are mentioned in the *Rgveda* . *Dahi* (curd) was such a product that was used as an essential offering to god and was also an ingredient for the preparation of *soma*juice. Different methods of curdling of milk are described in different texts. For instance, the method described in*Taittiriya Samhita* mentions the use of whey and rice grains for curdling of milk.

Post-Vedic Period (600BC to 100CE)

In this period, two new preparations known as vinegar and liquor from bread were added by adopting new technologies. Later on some other drinks were also prepared by using fermentation technologies. Various developments took place in the field of fermentation technology during the post-Vedic period. Some of these developments are described below:

- Now, a large number of base materials were used instead of a few. Some common substances that began to be used as base material were rice and barley among cereals; grape, palm, mango, wood-apple and sugarcane, etc, among fruits; madhuka (Bassia latifolia) and jasmine among flowers and barks ofmesasrngi , putraka, etc.
- Earlier honey was mostly used as a ferment, but later on treacle (AS. II.25) also began to be used and became very popular in a short span of time.
- A new period began with the employment of fermented products in therapeutic medicines and chemical practices.
- Auxiliary (sambhara) (AS .II.25): It was a compound that served dual purposes. On the one hand it was used as ferment, on the other it was also used to add flavour and taste to liquor. This compound was prepared by using cinnamon, Plumbago zeylanica , Emblia ribes and gaza-pippali (elephant pepper plant). From sambhara a special liquid known as Sambhariki was also prepared.

Technologies connected with the preparation of *beer* (*sura*)

The Buddhist texts mention five different types of *sura,* but the technologies involved in the preparation of these are not discussed. Some texts mention that germinated cereals and hulled fruit drenched in water dried, ground and given the shape of a cake was known as bread. Later on brewing was done by soaking it in water which in turn was fermented by adding some yeast to that solution.

Technologies connected with the preparation of *vinegar* (*amla*)

According to Kautilya, there were two types of *amla* preparations. One method involved the use of fruit juice and*phalamla* , while the other was derived from molasses. Its preparation was done for a longer time so that the product became sour. Kautilya further states that these were home-made products.

Medaka (spiced rice beer), *Prasanna* (spiced barley or wheat beer), *Asava* (sugarcane beer), *Arista* (medical tincture), etc, were some of the most popular drinks of that period. Different varieties of fermented liquor provide us the knowledge of different fermentation processes developed during that period.

Period of Scientific and Technical Literature (100 CE onwards)

Fermentation technology advanced in the field of medical science and chemical and alchemical practices. A large number of digesting methods now came into existence. Some of these digesting methods were:

- The fermenting vessel was deposited into the earth and sometimes the vessel was covered with horse dung.
- The fermenting vessel was placed on a heap of grains.
- The fermenting vessel was exposed to the sun.
- Fumigation of the desired substance was done in a pot.

Some new recipes for fermentation were formulated. One such recipe is described in *Rasopanisat* (*RP* . XV 251-253). The recipe is:

"The best ones among the five classes of bulbous plant with latex pounded along with the grains of rice of *Kodrava* (*Paspaln scrobiculatum*), and products of plant *madana* (emetic nut), pasted with whey of buffalo or cow-milk and kept in a closed bowl and another (?). The closed vessel is then placed in the sun. The acidic residue thus obtained is*kinva* (*yeast*)."

The alcoholic fermentation and acetous fermentation began to be used for several purposes now by changing the base material or by slightly modifying the digesting methods. Base material was known as *yoni* and the fermented produces were known as *tri-yoni, dvi-yoni,* etc. The juicy produce that was obtained from the fermented smash was known as *Jata-rasa* . The liquid obtained from the boiled substance was also used for the processes of fermentation.

Fermentation products, arista, asava and sura, were considered potent medicinal drugs. Similarly, aranala, cukra, dhanyamala, kanjika, sauviraka, sukta, tusambu, etc, were not only used for medicinal purposes but also used for washing of metals and minerals and for different mercurial operations.

Practices involved in spirituous (alcoholic) fermentation (*SS.* Ci . 10.6-8)

Arista : Powder of certain selected drugs of desired medicinal properties as well as fermenting properties mixed with water, iron powder and treacle was used for the preparation of *arista* . Water taken was about half of the weight of the powder while iron powder and treacle taken were 28 ½ times the weight of water. All of these ingredients were placed in an earthen vessel. The earthen vessel was coated with a mixture of honey, butter and powdered long-pepper plant from inside.

Now the sealed earthen pot was placed in a mass of barley for at least seven nights for fermentation to take place.

Asava : A mixture of either cold alkaline water or sesamum and juice of sugarcane was taken for the preparation of*asava* . These were taken in the ratio of 2: 3 respectively. The process of fermentation was the same as that of*arista* .

Sura : In the preparation of *Sura* a decoction of a mixture of desirable drugs was taken with the ferment. This preparation was also similar to that of *arista* and *asava* .

Preparation of special *Sura* for purification (*SS.* Su . 44. 29-34)

A mixture of decoctions of prescribed plants and sugar was taken in the ratio of 2: 3 respectively. This mixture was then boiled over fire. After boiling the mixture, the solution was kept for cooling. Now, the solution was placed in an earthen jar for about a month. The time period was taken in such a way that half of this period fell in the winter and the other half in the anterior or posterior of winter. The fermentation process took place during this period. Likewise, alkaline water, cow's urine and some other substances were also used for similar preparations.

Not only references to these medicated liquors but a number of other fermented liquors with their respective medicinal values are found in *Caraka* and *Susruta-Samhita* .

PRACTICES INVOLVED IN ACETIC FERMENTATION

Some common practices are described below:

Aranala: Husked uncooked wheat or any cereal was taken as a base material with any ferment and the mixture was heated for about a week by solar radiation.

Cukra: Bulging root, fruit of the desired plant was taken with alcoholic sweet juicy whey and the mixture was deposited and purified in a heap of paddy for about three nights.

Sukta: A mixture of treacle, honey, fermented rice water, whey was used in the preparation of *sukta* . This mixture was placed in an earthen pot and the pot was placed on heaps of paddy for three nights in the summer season.

Kanjika: Boiled millet or barley was used as a base material. Different plants were used as a ferment medium. It could also be prepared by using boiled rice.

The above facts clearly show that the history of fermentation technology in India from the ancient times onwards includes a variety of techniques which resulted in a rich repertoire of alcoholic beverages and allied products. What is important is that fire and heat were rarely employed for the process of fermentation.

Traditional Fermentation 101 and a Recipe: Quick & Easy Gingered Carrot

In the long span of history before refrigeration, fermenting was the primary method of preserving food for long periods, ensuring that families would have nourishment in lean times or over cold winters when fields lay fallow. It's a practice found in traditional cultures around the world: Germany is well-known for sauerkraut but it is actually an ancient food that the Romans prized both for taste and medicinal qualities (and probably imported to that region of Europe now called Germany). Korean and Japanese meals, even today, are not complete without some sort of pickled vegetable served on the side (kimchi or umeboshi plums, anyone?) to which they ascribe great health-giving properties. American pioneers made the most of summer harvests by fermenting corn, cucumbers and even watermelon rinds into relish and pickles,turning ripe tomatoes into ketchup. Honestly, just about every kind of food out there can be fermented, although you might better match them up with synonyms like curing (meats) and culturing (dairy). Today I'm focusing on fermenting vegetables.

All right, you might be asking, so why is traditional fermentation better? To quote Sally Fallon from my kitchen bible, *Nourishing Traditions.*:

"...preservation of vegetables and fruits by the process of lacto-fermentation has numerous advantages beyond those of simple preservation. The proliferation oflactobacilli in fermented vegetables enhances their digestibility and increases vitamin levels. These beneficial organisms produce numerous helpful enzymes as well as antibiotic and anticarcinogenic substances. Their main by-product, lactic acid, not only keeps vegetables and fruits in a state of perfect preservation but also promotes the growth of healthy flora throughout the intestine."

Traditional fermentation is not a quick process (although preparation time is short), but leave it to the 20th century and its penchant for speeding everything up, to replace the old ways with quick-fermenting vinegar which, unfortunately, bypasses the nutrient-building that happens with a slow, traditional ferment. As we became more divorced from nature (again, largely a 20th century invention) we also became micro-phobic, desiring to eradicate any bacteria from our foods. The process of widespread pasteurization killed off harmful bacteria but along with it killed the beneficial bacteria that would have likely taken care of the bad guys themselves, while also making themselves available to populate our guts for better digestion and overall health. The result is what I call a "dead" food – pasteurization literally kills the life in our food. I believe (and I'm not the only one) that this now-common practice has led to the proliferation of food allergies, intestinal disorders like Irritable Bowel Syndrome and Crohn's disease, and the epidemic rise of autism which, along with schizophrenia and ADHD, has been strongly linked to gut disorder.

A few years ago I started checking labels of the fermented foods I bought, trying to find brands that use traditional ingredients like salt, sugar, whey or lemon juice. They are out there if you *really* look, but eventually I decided I'd just learn how to make those foods myself. It's far cheaper and more satisfying.

Fermentation is a craft, one of those domestic arts that was nearly lost but is being revived, thanks largely to the internet and the traditional foods community, led by websites such as Real Food Media, Nourished Kitchen, Cheeseslave and

Nourishing Days (and, now, Maggie's Nest!). Books like *Nourishing Traditions* and *Wild Fermentation* also champion the effort to bring this ancient kitchen wisdom into the 21st century.

Today's recipe, modified slightly from *Nourishing Traditions*, is touted as their favorite introduction to traditionally fermented foods: "the taste is delicious; and the sweetness of the carrots neutralizes the acidity that some people find disagreeable when they are first introduced to lacto-fermented vegetables. Ginger carrots go well with rich foods and spicy meats."

Benefits of Traditional Fermented

All around the world, fermented foods and beverages are part of the human diet. In some places they make up a minor 5% of daily intake, while in others their role can be as substantial as 40%.

Using native knowledge of locally available raw materials from plant or animal sources, people across the globe produce this type of food and drink either naturally or by adding starter cultures that contain micro-organisms. Micro-organisms transform these raw materials both biochemically (i.e., the nutrients) andorganoleptically (i.e., the taste/texture/odour) into edible products that are culturally acceptable to the maker and consumer.

Fermented foods can be fried, boiled or candied, or consumed in curries, stews, side dishes, pickles, confectionery, salads, soups and desserts. They can be in the form of pastes, seasonings, condiments, masticators, and even colourants. Fermented drinks can be either alcoholic (such as beer and wine) or non-alcoholic, like butter milk, certain teas, or things that contain vinegar.

However, though most fermented foods have health-promoting benefits, their global consumption is declining as traditional food systems give way to the influence of a western diet and fast foods.

A world of therapeutic food

Many of the fermented products consumed by different ethnic groups have therapeutic values. Some of the most widely known are fermented milks (i.e., yoghurt, curds). Containing high concentrations of pro-biotic bacteria, these can lower your cholesterol level.

One fermented food that has received much attention is the viscous, acidic, mildly alcoholic milk beverage produced by fermentation of milk with a particular grain in Eastern European and Middle Eastern countries. *Kefir* is easily digested and provides the human body with beneficial micro-organisms that contribute to a healthy immune system — a boon to patients suffering from AIDS, chronic fatigue syndrome, herpes and cancer. Traditionally it has been used for the treatment of tuberculosis and cancer. Another naturally fermented dairy product, *koumiss*, found in the Caucasus region, is used in the treatment of pulmonary tuberculosis.

Other widely-eaten types of health-boosting sustenance include fermented soybean products found around Asia. Antioxidant and other properties are reported to exist in foods such as *natto*, a sticky dish high in protein and popular at breakfast in Japan, that may help prevent people from having brain haemorrhages. *Natto* is also rich in vitamin K2, which stimulates the formation of bones and might help to prevent osteoporosis in older people. Similarly, consumption of Indonesia's *tempe* reduces cholesterol levels and, like China's *douchi*, lowers high blood pressure.

Korea's famous accompaniment, *kimchi*, may take the prize for most benefits however, reportedly helping to prevent constipation and colon cancer and reduce serum cholesterol, as well as possessing anti-stress effects and the ability to ameliorate depression, osteoarthritis, liver disease, obesity and atherosclerosis.

In the Himalayas, a fermented leafy vegetable product called *gundruk* and a fermented radish tap-root (*sinki*) have large amounts of lactic acids, ascorbic acid, carotene and dietary fibre, which have anti-carcinogenic effects. In order to gain strength, ailing persons and post-natal women in the Himalayas consume *bhaati jaanr* extract (a fermented rice food-beverage) and *kodo ko jaanr* (a fermented finger millet product) due to their high calorie content.

Fermentation's magic

One of the important outcomes of food fermentation is the enrichment of food with essential amino acids, vitamins, mineral and bioactive compounds. For example, during tempe fermentation, vitamins like niacin and riboflavin are increased by using the starter culture *Rhizopus oligosporus* . Similarly, thiamine and riboflavin increase during fermentation of *idli*, the fermented rice and black-gram (a type of legume) product of India and Sri Lanka.

These clever processes developed by humans to extract additional nutritional value from food, are not only found in Asia. *Pulque* — one of the oldest alcoholic beverages prepared from the juices of cactus plants in Mexico — is rich in vitamins such as thiamine, riboflavin, niacin and biotin.

Another important reason to ferment foods is to coax micro-organisms into producing enzymes that also provide very useful services. These enzymes degrade anti-nutritive compounds and thereby make edible, with enhanced flavour and aroma, things that otherwise would be indigestible and/or unpalatable. Bitter varieties of cassava tubers contain a potentially poisonous substance that can be detoxified via lactic acid bacteria, as in *gari* and *fufu*, fermented cassava root products from Africa.

Many people suffer from lactose intolerance or lactose malabsorption, a condition that causes lactose, the principal carbohydrate of milk, to not be completely digested. The cultures used in making yoghurt and curds, contain substantial quantities of ß-D-galactosidase, something that is thought to help alleviate the symptoms of lactose malabsorption. Yoghurt, as a viscous food, may delay the stomach emptying and that way help lessen lactose intolerant symptoms.

HEALTHY, BUT NOT APPRECIATED

Although fermented foods are marketed globally as health foods, functional foods, therapeutic foods, nutraceutical foods or bio-foods, due to urbanization, changes in life-style, and the shifting from traditional food habits to commercial fast foods, the production and consumption of traditional fermented foods is in decline.

Even when fermented foods are being consumed, they are increasingly likely to be from packets in supermarkets, rather than directly from households or a local family run business. This is leading to fewer and fewer people possessing traditional knowledge of fermented foods and the ability to make these from scratch. Reliance on fewer providers of fermented foods is also leading to a decline in the biodiversity of micro-organisms, or 'microbial-biodiversity'.

Improving the Efficiency of Traditional Fermentation Industries

Most of the traditional food fermentation industries could be considered as small household or backyard industries with small capacity production and inefficient in the application of the fermentation technology process and know how. It is not that difficult to improve or upgrade the traditional fermentation industries with little capital input and application of GMP

The characteristics of traditional fermentation industries in Malaysia are:

1. They are small scale or small volume industries
2. There are a diversity of fermentation products produced mre towards the geographical location and the type of food substrate available in the area. Coastal areas may produce fermentation products from cincaluk, belacan, budu. Other more inland areas might be producing tempe, tapai and kicap fermentation
3. Level of technology utilized is very low and do not really changed with time and regarded as passive fermentation. Use of automation and mechanization is low
4. GMP is rarely applied. Lack of quality control and sanitary standards
5. Very low level of research or scientific input
6. Government industries supposed to support and upgrade these small and scattered industries are quite passive and do not really enter the playground to participate
7. Shelf life for some of the fermentation products are really short in case of tempe and tapai
8. Marketting of the traditional fermentation products considered poor
9. These industries do not really have professional consultants in fermentation who can advise them how to improve and standardize the quality of fermentations
10. Most of the traditional fermentations are Batch or fed batch fermentations
11. Operators of the traditional fermentation have poor or lack of understanding of the science of fermentation that be applied to their process

12. The fermentation industry is a mix of Solid substrate and liquid fermentation

In this write up we will only try to discuss one or two aspects of the traditional fermentation industries:

1. Increasing the volume of fermentation production
2. Improving the efficiency of the traditional fermentation industries with simple cheaper and alternative modes of fermentation

INCREASING THE VOLUME OF FERMENTATION

One of the most common characteristics of traditional fermentation here is that the fermentation process are often carried out in a large number of small fermentation vats (or earthen jars). This limiting step can be avoided by carrying the fermentation process in a large fermentation vessel. The most ideal container for the fermentation now would be adapting or modifying the new stainless steel water tanks. These tanks are easily available from hardwate shops. These tanks are sanitary and are easily cleaned and sterilize

Adding a simple motor to the top of the fermentor or stainless steel water tank will help improve the mixings and the fermentation process.

Increasing the Efficiency of the Fermentation Process

Improvement of the traditional fermentation process has always been the rate limiting steps in the fermentation process. To improve the fermentation process must be carried out in terms of time, quality and volume of the fermentation process.

When we talk about shortening the fermentation time by optimizing the various physical, chemical and microbiological as well as the engineering aspects of the fermentor operation. This initially requires some laboratory scale fermentation studies to optimize the various parameters.

The product of the fermentation should be of the highest quality and consistency. Standard operating procedures and good GMP or the basics of GMP should be adapted or modified for use at the level of cottage industries fermentation. The HAACP procedures should also be adapted in its essence to be applied at the small scale fermentation.

The main problem of traditional fermentation industries in this country it has always been so successful in maintaining the traditional way of thinking and making fermentations. There is lack of tendency to understand the process, to increase the input of science and technology to the fermentation production system. Days of fermenting in plastic pails and porcelain vats are numbered.......

In my observations the fermentation entrepreneurs still practise the art of fermentation handed down from the fathers and refusing to optimise or even experiment to improve their fermentation process. There is poor monitoring of the fermentation process in terms of temperature, ph , sugar in the traditional fermentation industries. This has led to inconsistency in the fermentation products produced. A little investment in these instruments would increase the efficiency of their fermentation.

ADVANCES IN TRADITIONAL FERMENTATION

Fermentation technology has been a long-favored biological process due to its simplicity of reaction, high specificity, low costs, and versatility of application. Modern industries have complemented the basic principle of fermentation with advances in techniques and genetic engineering to extend its application to the production of a range of biochemicals, biomolecules, and biofuels. Fermentation's applications are rapidly breaching the confines of food stuff such as wine, cheese, bread, and beer, into high-value chemicals, pharmaceutical ingredients, and food-related chemicals. Further, rising petrol prices and depleting fossil fuel reserves have made a robust case for low-cost and simple fermentation processes for producing drop-in chemical replacements. There has been a spurt in research efforts, both in the corporate and academic sector, driven largely by fermentation applications within the energy and utilities, specialty chemicals, and pharmaceuticals industries. Energy - or more specifically, biofuels - has evinced considerable interest in applying fermentative techniques to produce cost-competitive bioethanol.

While the uses of fermentation techniques are many in modern industrial processes, there are several challenges to overcome before implementing them in large-scale commercial applications. One of the stand-out challenges is that fermentation is a slow, natural bioprocess, which means that time-scale of operations is often quite long. Companies also have to contend with the low yields and poor process economics in large-scale production. Although fermentation operating costs are lower than full-scale chemical operations, the costs of raw materials and fermentable sugars remain a challenge for most companies. In addition, specially designed enzymes and microbes with specific biosynthetic pathways are also expensive. Therefore, the end products derived will be hard pressed to be cost competitive with other petrochemical-derived alternatives.

To stave off these challenges, companies are increasingly developing robust proprietary microorganisms that can ferment the substrates to produce the desired products. Consequently, companies that are able to provide a solution that combines excellence in microbe technologies with superiority in process technologies will be better equipped to explore new application avenues. Importantly, companies that can take advantage of fermentation's unique bio-capabilities, while still remaining economically feasible, will find themselves operating in a market with expansive possibilities. Many companies have begun engineering innovative enzymes, microbes, and bacterial strains that can help produce biomolecules in bulk. Apart from genetic techniques, some of the recent developments that continue to be explored include process-driven technologies such as solid-state fermentation, consolidated bioprocessing, ultrasonication, syngas fermentation, and dark fermentation. These developments are indicative of the huge potential of fermentation as a low-cost and flexible method to produce a range of end products.

Solid state fermentation for the production of industrial enzymes

Enzymes are among the most important products obtained for human needs through microbial sources. A large number of industrial processes in the areas of industrial, environmental and food biotechnology utilize enzymes at some stage or the other. Current developments in biotechnology are yielding new applications for enzymes. Solid state fermentation (SSF) holds tremendous potential for the production of enzymes. It can be of special interest in those processes where the crude fermented products may be used directly as enzyme sources. This review focuses on the production of various industrial enzymes by SSF processes. Following a brief discussion of the micro-organisms and the substrates used in SSF systems, and aspects of the design of fermenter and the factors affecting production of enzymes, an illustrative survey is presented on various individual groups of enzymes such as cellulolytic, pectinolytic, ligninolytic, amylolytic and lipolytic enzymes, etc.

Solid state fermentation (SSF) holds tremendous potential for the production of enzymes. It can be of special interest in those processes where the crude fermented product may be used directly as the enzyme source. In addition to the conventional applications in food and fermentation industries, microbial enzymes have attained significant role in biotransformations involving organic solvent media, mainly for bioactive compounds. Table 1 lists some of the possible applications of the enzymes produced in SSF systems. This system offers numerous advantages over submerged fermentation (SmF) system, including high volumetric productivity, relatively

higher concentration of the products, less effluent generation, requirement for simple fermentation equipments, etc..

Microorganisms used for the production of enzymes in solid state fermentation systems

A large number of microorganisms, including bacteria, yeast and fungi produce different groups of enzymes. Selection of a particular strain, however, remains a tedious task, especially when commercially competent enzyme yields are to be achieved. For example, it has been reported that while a strain of *Aspergillus niger* produced 19 types of enzymes, a -amylase was being produced by as many as 28 microbial cultures. Thus, the selection of a suitable strain for the required purpose depends upon a number of factors, in particular upon the nature of the substrate and environmental conditions. Generally, hydrolytic enzymes, e.g. cellulases, xylanases, pectinases, etc. are produced by fungal cultures, since such enzymes are used in nature by fungi for their growth. *Trichoderma* spp. and *Aspergillus* spp. have most widely been used for these enzymes. Amylolytic enzymes too are commonly produced by filamentous fungi and the preferred strains belong to the species of *Aspergillus* and *Rhizopus*. Although commercial production of amylases is carried out using both fungal and bacterial cultures, bacterial a -amylase is generally preferred for starch liquefaction due to its high temperature stability. In order to achieve high productivity with less production cost, apparently, genetically modified strains would hold the key to enzyme production.

The selection of a substrate for enzyme production in a SSF process depends upon several factors, mainly related with cost and availability of the substrate, and thus may involve screening of several agro-industrial residues. In a SSF process, the solid substrate not only supplies the nutrients to the microbial culture growing in it but also serves as an anchorage for the cells. The substrate that provides all the needed nutrients to the microorganisms growing in it should be considered as the ideal substrate. However, some of the nutrients may be available in sub-optimal concentrations, or even absent in the substrates. In such cases, it would become necessary to supplement them externally with these. It has also been a practice to pre-treat (chemically or mechanically) some of the substrates before using in SSF processes (e.g. ligno-cellulose), thereby making them more easily accessible for microbial growth.

Among the several factors that are important for microbial growth and enzyme production using a particular substrate, particle size and moisture level/water activity are the most critical. Generally, smaller substrate particles provide larger surface area for microbial attack and, thus, are a desirable factor. However, too small a substrate particle may result in substrate agumulation, which may interfere with microbial respiration/ aeration, and therefore result in poor growth. In contrast, larger particles provide better respiration/aeration efficiency (due to increased inter-particle space), but provide limited surface for microbial attack. This necessitates a compromised particle size for a particular process.

SSF processes are distinct from submerged fermentation (SmF) culturing, since microbial growth and product formation occurs at or near the surface of the solid substrate particle having low moisture contents. Thus, it is crucial to provide an optimized water content, and control the water activity (*a*w) of the fermenting substrate—for, the availability of water in lower or higher concentrations affects microbial activity adversely. Moreover, water has profound impact on the physico-chemical properties of the solids and this, in turn, affects the overall process productivity.

Aspects of design of fermenter for enzyme production in solid state fermentation systems

Over the years, different types of fermenters (bioreactors) have been employed for various purposes in SSF systems. Pandey reviewed the aspects of design of fermenter in SSF processes. Laboratory studies are generally carried out in Erlenmeyer flasks, beakers, petri dishes, roux bottles, jars and glass tubes (as column fermenter). Large-scale fermentation has been carried out in tray-, drum- or deep-trough type fermenters. The development of a simple and practical fermenter with automation, is yet to be achieved for the SSF processes.

Factors affecting enzyme production in solid state fermentation systems

The major factors that affect microbial synthesis of enzymes in a SSF system include: selection of a suitable substrate and microorganism; pre-treatment of the substrate; particle size (inter-particle space and surface area) of the substrate; water content and *a*w of the substrate; relative humidity; type and size of the inoculum; control of temperature of fermenting matter/removal of metabolic heat; period of cultivation; maintenance of uniformity in the environment of SSF system, and the gaseous atmos-phere, i.e. oxygen consumption rate and carbon dioxide evolution rate.

Enzymes produced by solid state fermentation processes

Ideally, almost all the known microbial enzymes can be produced under SSF systems. Literature survey reveals that much work has been carried out on the production of enzymes of industrial importance, like proteases, cellulases, ligninases, xylanases, pectinases, amylases, glucoamylases, etc.; and attempts are also being made to study SSF processes for the production of inulinases, phytases, tannases, phenolic acid esterases, microbial rennets, aryl-alcohol oxidases, oligosaccharide oxidases, tannin acyl hydrolase, a -L-arabinofuranosidase, etc. using SSF systems . In the following sections, a brief account of production on various enzymes in SSF systems is discussed.

Cellulases, Xylanases and Xylosidases

Cellulases are a complex enzyme system, comprising endo-1,4-b -D-glucanase (EC-3.2.1.4), exo-1,4-b -glucanase (exocellobiohydrolase, EC-3.2.1.91) and b -D-glucosidase (b -D-glucoside glucanhydrolase, EC-3.2.1.21). These enzymes, together with other related enzymes, viz.hemicellulases and pectinases, are among the most important group of enzymes that are employed in the processing of ligno-cellulosic materials for the production of feed, fuel, and chemical feedstocks. Cellulases and xylanases (endo-1,4-b -D-xylanase, EC-3.2.1.8) however find applications in several other areas, like in textile industry for fibre treatment and in retting process. Xylanases find specific application in jute fibre upgradation also.

Currently, industrial demand for cellulases is being met by production methods using submerged fermentation (SmF) processes, employing generally genetically modified strains of *Trichoderma*. The cost of production in SmF systems is however high and it is uneconomical to use them in many of the aforesaid processes. This therefore necessitate reduction in production cost by deploying alternative methods, for example the SSF systems.

Tengerdy compared cellulase production in SmF and SSF systems. While the production cost in the crude fermentation by SmF was about $ 20/kg, by SSF it was only $ 0.2/kg if *in situ*fermentation was used. The enzyme in SSF crude product was concentrated; thus it could be used directly in such agro-biotechnological applications as silage or feed additive, ligno-cellulosic hydrolysis, and natural fibre (e.g. jute) processing. A number of reports have appeared on microbial cellulase production in recent years . Nigam and Singh have reviewed processing of agricultural wastes in SSF systems for cellulolytic enzyme production. They argued that with the appropriate technology, improved bioreactor design, and operation controls; SSF may become a competitive method for the production of cellulases. They also enumerated advantages of cellulase production together with the factors affecting the cellulase production in SSF systems.

In a recent study on the ligninolytic system of *Cerrena unicolor* 062 – a higher basidiomycete – upon supplementation of the medium with carbon sources and phenolic compounds in SSF system, it was observed that the growth of *C. unicolor* 062 could be regulated by the exogenous addition of these compounds. The efficiencies of the degradation of cellulose and lignin were dependent on the nature and concentration of the compounds added. Sun *et al.* developed a novel fed-batch SSF process for cellulase production which could overcome the problems associated with high initial nutrients concentration while retaining advantages from the high total effective salt concentration.

There are several reports describing co-culturing of two cultures for enhanced enzyme production. Gupte and Madamwar cultivated two strains of *Aspergillus ellipticus* and *A. fumigatus* and reported improved hydrolytic and b -glucosidase activities compared to when they were used separately using SSF system, improved enzyme titres were achieved by Kanotra and Mathur when a mutant of *Trichoderma reesei* was co-cultured with a strain of *Pleurotus sajor-caju* with wheat straw as the substrate. However, the media constituents too play an important role in mixed culturing. Gutierrez-Correa and Tengerdy reported that single culture of *T. reesei* and *Aspergillus phoenicus*, when supplemented with inorganic nitrogen source, produced similar xylanase levels as mixed cultures. However, when the fermentation medium was supplemented with soy meal, 35–45% more xylanase (than the single culture) was produced by these cultures.

In a significant finding, Smits *et al.* reported that glucosamine level of the fungi in liquid culture could not be used to estimate the biomass contents in SSF. They studied the SSF of wheat bran by *T. reesei* and reported that using glucosamine, correlation between the fungal growth and respiration kinetics could only partly be described with the linear growth model of Pirt. A decline in O2 consumption rate (OCR) and CO2 evolution rate (CER) started the moment glucosamine was 50% of its maximum value. After the glucosamine level reached its maximum, OCR and CER still continued to decrease.

A pan bioreactor, requiring a small capital investment, was developed for SSF of wheat straw. High yields of complete cellulase system were obtained in comparison to those in the SmF. A complete cellulase system is defined as one in which the ratio of the b -glucosidase activity to filter paper activity in the enzyme solution is close to 1.0. The prototype pan bioreactor however required further improvements so that optimum quantity of the substrate could be fermented to obtain high yields of complete cellulase system per unit space.

Although xylanases produced by fungi, yeast and bacteria, filamentous fungi are preferred for commercial production as the levels of the enzyme produced by fungal cultures are higher than those obtained from yeast or bacteria. In many microorganisms, xylanase activity has generally been found in association with cellulases, b -glucosidase or other enzymes, although there are many reports that have described in SSF systems, production of cellulase-free and other enzymes-free xylanase. Haltrich *et al.* reviewed the different factors that influence xylanase production by fungi. In view of the considerable commercial importance of enzymes, it was emphasized that efforts should be directed towards enhanced enzyme production with reduced associated costs.

Archana and Satyanarayana described a SSF process for the production of thermostable xylanase by thermophilic *Bacillus licheniformis*. Enzyme production was 22-fold higher in SSF system than in SmF system. Cai *et al.* also reported production of a thermostable xylanase in SSF system. Enzyme produced in SSF system was more thermostable than in SmF system. Dunlop *et al.* described a bacterium, isolated from wood compost, producing xylanase that was active at 80°C. Jain too described a SSF process for the production of xylanase by thermostable*Melanocarpus albomyces*.

Alam *et al.* using SSF process, isolated a thermostable cellulase-free xylanase produced by *T. lanuginosa*. Addition of 0.7% xylan induced enzyme production to an extent of 28%. The enzyme was stable at 70°C. A thermostable xylanase preparation from *Humicola* sp. showed the temperature optima at 75°C (ref. 87). Srivastava reported a xylanase from *Thermomonospora*sp., which was stable at 80°C. Tuohy and Coughlan compared thermostable xylanase production on various substrates by a strain of *Talaromyces emersonii* in liquid culture and SSF systems. The latter showed higher enzyme activity compared to former, but liquid culture resulted in greater yields (U/g substrate).

Several authors have compared the performance of various microbial strains, grown on different substrates (individual or in combination) and reported varying results. Wiacek-Zychlinska *et al.* compared xylanase production by *C. globosum* and *A. niger* on four different substrates. Although activities obtained by *A. niger* were higher than those from the other microbial cultures, but high-spore production by the *A. niger* strain could result in problems for a pilot plant or large-scale process.

In order to achieve improved enzymes titre, it is generally a common practice to pre-treat cellulosic or ligno-cellulosic substrates before using them in SSF systems. Pre-treatment may be by physical processes or chemical processes. Pre-treatment of palm oil mill waste, however, did not affect xylanase production.

b -xylosidase is another important enzyme used in textile industry. A b -xylosidase (EC-3.2.1.37) was produced by *A. awamori* K4 in SSF system on wheat bran, which was used for transxylosylation reactions. There are other reports as well describing the production of b -xylosidase in SSF systems.

Ligninases

Lignin is a three-dimensional phenylpropanoid polymer which is considerably resistant to microbial degradation in comparison to polysaccharides and other naturally occurring biopolymers. Biological delignification by SSF processes using microbial cultures producing ligninolytic enzymes – the ligninases – can have applications in delignification of ligno-cellulosic materials , which can be used as the feedstock for the production of biofuels or in paper industry or as animal feedstuff. These may also be used in pulp bleaching, paper mill wastewater detoxification, pollutant degradation, or conversion of lignin into valuable chemicals.

Lignin peroxidase (LiP, EC-1.11.1.7), manganese peroxidase (MnP, EC-1.11.1.13) and laccase (EC-1.10.3.2) are the most important lignin-modifying enzymes. LiP and MnP are heme-containing glycoproteins requiring hydrogen peroxide as an oxidant. LiP oxidizes nonphenolic lignin structures by abstracting one electron and generating cation radicals, which are then decomposed chemically. MnP oxidizes Mn(II) to Mn(III), which then oxidizes phenolic compounds to phenoxy radicals. This leads to the decomposition of the lignin substructure. Laccase, a copper containing oxidase, utilizes molecular oxygen as the oxidant and oxidizes phenolic components to phenoxy radicals.

Literature survey shows that a number of microorganisms produce ligninases, but white-rot fungi generally show the most desirable qualities, in particular *Pleurotus* species and *Phanerochaete chrysosporium* are the most widely studied.

Wheat straw was used for cultivating several fungal strains to produce laccase, Li-peroxidase, and Mn-peroxidase. Several authors have used bagasse also. Homolka *et al.* studied laccase production from three strains of *Pleurotus* sp. (obtained after protoplast regeneration of the control strain). While two strains showed significantly higher laccase activity, one strain showed lower activity. The rate of mineralization of C-lignin in SSF system by the latter and the control strain were almost the same, but it was higher than that of the other two strains. C-lignin in SSF of wheat straw was also used by Camarero *et al.* for studying Mn-mediated lignin degradation by four strains of *Pleurotus* sp., and comparing with by *P. chrysosporium*. At the end of the incubation period, strains of *Pleurotus* sp. acquired higher delignification values than *P. chrysosporium*. All the species of genus *Pleurotus*, studied so far, produce Mn-peroxidase, laccase, and aryl-alcohol-oxidase (EC-1.1.3.13).

Dombrovskaya and Kostyshin studied the effects of different ionic nature surfactants on ligninolytic enzyme complexes of the white-rot fungi in SSF processes.

The cationic surfactant, ethonium, enhanced the laccase and Mn-peroxidase activity by 1.8 fold and 1.6 fold, respectively for *P. floridae.* Kerem and Hadar studied the effects of Mn on the production of ligninolytic enzyme complexes of *P. ostreatus* in a chemically defined SSF system. Laccase, Mn-peroxidase, and catalase (EC-1.11.1.6) activities, and H2O2 production were all affected by Mn levels.

Laplante and Chahal compared ligninase production in SmF system and SSF system using a culture of *P. chrysosporium* ATCC 24725. Higher yields of ligninases, especially laccase and Mn-peroxidase, were obtained in SSF system. Kerem *et al.* compared the ligninolytic activity of a strain of *P. chrsosporium* BKM with *P. ostreatus* Florida f16. The former grew vigorously resulting in rapid, non-selective degradation of 55% of the organic components of the cotton stalks within 15 days. *P. ostreatus* grew more slowly with obvious selectivity for lignin degradation, resulting in the degradation of only 20% or the organic matter in 30 days.

Proteases

Proteolytic enzymes account for nearly 60% of the industrial market in the world. They find application in a number of biotechnological processes, viz. in food processing and pharmaceuticals, leather industry, detergent industry, etc. Recently, Mitra *et al.* reviewed production of proteolytic enzymes in SSF systems. From their viewpoint, proteases produced by SSF processes have greater economic feasibility.

In recent years, there have been increasing attempts to produce different types of proteases (acid, neutral, alkaline) through SSF route, using agro-industrial residues. It is interesting to note that although a number of substrates have been employed for cultivating different microorganisms, wheat bran has been the preferred choice in most of the studies. Malathi and Chakraborty evaluated a number of carbon sources (brans) for alkaline protease production and reported wheat bran to be the best for cultivation of *A. flavus* IMI 327634. Studies were carried out to compare alkaline protease production in SmF systems and SSF systems. The total protease activity present in one-gram bran (SSF) was equivalent to 100-ml broth (SmF). A repeated batch mode SSF process was described for alkaline protease production in which polyurethane was used as the inert solid support. A thermostable alkaline protease was reported to be produced by a novel *Pseudomonas* sp. in SSF system. A process has been developed at CLRI, Chennai (India), for the commercial production of an alkaline protease (Clarizyme) which was produced by SSF of wheat bran using a strain of *A. flavus*.

A new strain of *A. niger* Tieghem 331221 produced large quantities of an extra-cellular acid protease when grown in SSF system using wheat bran as the sole substrate. Various C-sources inhibited protease synthesis, indicating the presence of catabolic repression of protease biosynthesis. The enzyme showed potential for usage as a bating agent. Ikasari and Mitchell used rice bran for acid protease synthesis by a strain of *R. oligospora*. They observed that although the enzyme showed optimum activity at pH 4, a leaching solution of pH 7 gave the optimum recovery of the enzyme from the fermented matter. They made stepwise changes in the gas environment and temperature during SSF process to mimic those changes which arose during SSF due to mass and heat transfer limitations. It was observed that a decrease of O2 concentration from 21% to 0.5% did not alter protease production. Yaoxing *et al.* carried out SSF of wheat bran with a strain of *A. niger* QX 1066 for acid-resistant protease. High enzyme activities were obtained in a medium containing high carbon and low nitrogen content. Addition of a suitable phosphate in the medium further improved the enzyme titres. Villegas *et al.* studied the effects of O2 and CO2 partial pressure on acid protease production by a strain of *A. niger* ANH-15 in SSF of wheat barn. Results showed a direct relationship between pressure drop, production of CO2, and temperature increase. Acid protease production increased when the gas had 4% CO2 (v/v), and it was directly related with the fungus metabolic activity as represented by the total CO2 evolved.

Germano *et al.* used a strain of *P. citrinum* for serine protease production using agro-industrial residues. The strain also exhibited lipase activity. Datta used aspen wood for the production of protease from the fungal strain of *P. chrysosporium* BKM-F-1767. Study of this enzyme's characteristics showed that this protease had properties of aspartate-type protease as well as of thiol-type protease.

Lipases

Fat splitting has been completely revolutionized by the introduction of lipases (EC-3.1.1.3) into the industrial arena. The conventional physico-chemical means of lipolysis have now been undershadowed by the biocatalysis using microbial lipases. Lipases have a wide array of industrial applications in the production and processing of detergents, oils, fats and dairy-products.

In addition, they are also used in the preparation of therapeutic agents.

Until recently, SmF was in vogue for microbial lipase production. However, in recent years the shift has been towards the study and development of lipase production in SSF system. Beuchat investigated SSF of peanut press-cake using *Neurospora sitophila* and *Rhizopus oligosporus*. Rivera-Munoz *et al.* compared SmF systems and SSF systems for lipase production using several filamentous fungi. Enzyme titres by SSF processes were higher and stable. Among the tested microbial strains, *P. candidum*, *P. camembertii,* and *M. miehei* proved the best for lipase production.

Benjamin and Pandey and Benjamin cultivated *Candida rugosa* on coconut oil cake for lipase production using SSF and SmF systems. Enzyme yields were higher in the former. Several carbon sources – individually and in combinations – were tested for their efficiency to produce lipases. Raw cake supported the growth and lipase synthesis by the yeast culture. However, supplementation with additional C- and N-sources increased enzyme titres. In contrast to this, however, Ohnishi *et al.* reported less lipase production from *A. oryzae* using SSF compared to SmF where high enzyme yields were obtained. Yet, in another comparative study on lipase production in SmF and SSF systems, Christen *et al.* observed a 5-fold increase in lipase productivity in SSF system.

Bhusan *et al.* reported lipase production in SSF system from an alkalophilic yeast strain belonging to *Candida* sp. Rice bran and wheat bran, oiled with different concentrations of rice bran oil were used as the substrate. Rice bran supplemented with oil gave higher lipase yields. Ortiz-Vazquez *et al.* and Granados-Baeza *et al.* used wheat bran for cultivating the strains of *P. candidum*. They designed an enzyme-recovery procedure and reported that 0.01 M NaCl was adequate to recover enzyme from the fermented matter.

Pectinases

Studies have been conducted on comparative production of pectinases in systems of SmF and SSF. When the fermentation medium was supplemented with different carbon sources, like glucose, sucrose and galacturonic acid, polygalacturoanase (PG, EC-3.2.1.15) production by *A. niger* CH4 increased in SSF system but decreased in SmF system. Overall productivity by SSF was 18.8 and 4.9-fold higher for endo-PG and exo-PG, respectively, than those obtained by SmF. Minjares-Carranco *et al.* made physiological comparisons between pectinase-producing mutants of *A. niger* C28B25, adapted either to SmF or SSF. *A. niger* produced isozymes with difference in PG properties depending on the culture technique and strain used. The results also suggested that pleiotropic mutations of different kinds simultaneously affect the sporulation and enzymological patterns of each class of mutants.

Media acidity plays a significant role on pectinases' production by SSF processes. Cavalitto *et al.* and Hours *et al.* studied growth and pectinase production by *A. foetidus* and *A. awamori*, respectively in SSF systems at different media acidities. Both used wheat bran as the substrate. Results showed that higher the HCl concentration used, higher was the total pectolytic activity achieved. The low pH of the culture condition maintained asepsis during fermentation.

Apart from wheat bran, several other substrates have also been used for pectinase production in SSF system. These include coffee pulp , citrus waste , and apple pomace. Huerta *et al.* used bagasse as the inert substrate to produce PG in a 130 litres-packed bed fermenter by*A. niger* CH4 (they referred it as 'absorbed substrate fermentation'). They claimed that the process was an efficient one for PG production as well as an interesting model since the culture medium, water, nutrients and specific inducers could be varied depending on the concentrations required. Acuna-Arguelles *et al.* studied effect of water activity (*a*w) on exo-pectinase production by *A. niger* CH4 in SSF system. Sugar cane bagasse was used as the (inert) substrate and ethylene glycol was used as the water activity depressor. Results showed that although PG production decreased at low *a*w values, the activity was present even at as low as 0.90 *a*w values. The specific activity increased up to 4.5 fold by reducing the *a*w from 0.98 to 0.90.

Galactosidases

There has been considerable interest to produce a -galactosidase (EC-3.2.1.22) and b -galactosidase (EC-3.2.1.23) in SSF processes. Both these enzymes have applications in the pharmaceutical and food industries.

Cruz and Park reported production of a -galactosidase in SSF system and its application in the hydrolysis of galactooligosaccharides in soybean milk. Addition of soybean carbohydrate in the fermenting medium, using *A. oryzae*, was shown to induce enzyme production. Annunzaiato *et al.* carried out SSF of wheat bran for a -galactosidase production using a strain of *A. oryzae* QM 6737 with the aim of improving enzyme yields and lowering production costs. Enzyme yield increased 3 fold when soy flour or soybeans were used as the substrate, but no enzyme was produced using rice. Somiari and Balogh used a strain of *A. niger* for a -galactosidase production on wheat bran or rice bran. Srinivas *et al.* described the use of Plackett–Burman design for rapid screening of several nitrogen sources, growth/product promoters, minerals and enzyme inducers for the production of a -galactosidase by *A. niger* MRSS 234 in SSF.

In 1990, Wakamoto Pharma patented (two patents) the production of b -galactosidase in SSF systems. Strains of *Aspergillus* sp. and *Penicillium* sp. were used. Details have been provided in these patents by giving an example of the cultivation conditions and yields using a strain of *A. oryzae*. Enzyme preparation from *A. fonsecaeus*, which was cultivated on wheat bran , showed superior qualities than the other commercial preparation using a strain of *A. oryzae* and the enzyme was more suitable for biotechnological applications. Gonzalez and Monsan also used a strain of *A. fonsecaeus* for b -galactosidase production by SSF of wheat bran.

A thermostable b -galactosidase was reported from a thermophilic *Rhizomucor* sp. Enzyme activities by SSF were 9-fold more than by SmF processes. Strains of *Kluyveromyces* sp. have also been employed for b -galactosidase synthesis in SSF systems. Becherra and Siso cultivated *K. lactis* NRRL T-1140 on corn grits and wheat bran in SmF and SSF systems. They observed that change from liquid to solid state culturing did not promote b -galactosidase secretion by the yeast strain, though there were problems of drying of medium etc. in SSF. However, studies on production of b -galactosidase in SSF systems had already been published in 1995 (ref. 169).

Glutaminases

L-glutaminase is considered a potent anti-leukamic drug and has found application as a flavour-enhancing agent in food industry. In a maiden report, Prabhu and Chandrasekaran reported L-glutaminase production by SSF using marine *Vibrio costicola*. Polystyrene was used as the inert substrate. They also evaluated several organic substrates for their ability to produce glutaminases by SSF using the same strain. Among the tested materials, wheat bran and rice bran were found superior in comparison to saw dust, coconut oil cake, and groundnut cake. However, use of polystyrene as the substrate offered several advantages over organic substrtes. For example, leachate from polystyrene-SSF system was not only less viscous but also showed high specific activity of the enzyme.

Amylases

The amylase family of enzymes has been well characterized through the study of various microorganisms. Presence of two major classes of starch-degrading enzymes have been identified in the microorganisms, viz. *a* -amylase (endo-1,4-a -D-glucan glucohydrolase, EC-3.2.1.1) which randomly cleaves the 1,4-a -D-glucosidic linkages between the adjacent glucose units in linear amylose chain, and glucoamylase (synonym amyloglucosidase – also referred to as glucogenic enzyme, starch glucogenase, gamma amylase; exo-1,4-a -D-glucan glucanohydrolase, EC-3.2.1.3) which hydrolyses single glucose units from the nonreducing ends of amylose and amylopectin in a stepwise manner. Unlike a -amylase, most glucoamylases are also able to hydrolyse the 1,6-a -linkages at the branching points of amylopectin, although at a slower rate than 1,4-linkages.

Amylases and glucoamylases are produced by various microorganisms, including bacteria; fungi and yeast, but a single strain can produce both these enzymes as well. These enzymes have found applications in processed-food industry, fermentation technology, textile and paper industries, etc. Selvakumar *et al.* reviewed microbial synthesis of starch-saccharifying enzymes in solid cultures.

SSF has been employed to produce amylases. In a recent study, Ray *et al.* compared the production of b -amylase (EC-3.2.1.2) from starch waste by a hyper-amylolytic strain of *Bacillus megaterium* B6 mutant UN12 by SmF and SSF processes. The starchy wastes used as substrates were from arrowroot, arum, maize, potato, pulse, rice, rice husk, tamarind, kernel, cassava, water chestnut, wheat and wheat bran. Arum and wheat bran gave the highest yields.

Comparative studies on a -amylase production using different substrates have been studied as well. A new source of a -amylase was identified in *Pycnoporus sanguineus*. Cultivation of it in SSF system resulted in 4-fold higher enzyme production than in SmF system. Krishna and Chandrasekaran cultivated *Aeromonas caviae* (CBTK 185) on banana waste. The results indicated excellent scope for utilizing this strain and banana waste for commercial production of a -amylase by SSF. Sudo *et al.* compared acid-stable a -amylase production in SmF and SSF systems to ascertain as to why *A. kawachii* IFO 4308 produced larger amounts of acid-stable a -amylase in SSF system than in SmF system. Some of the attributes of SSF system were reported as the major reasons for higher enzyme production by SSF. A comparative study on SmF and SSF of inert substrate using a strain of *A. oryzae* CBS 125-59 also showed superiority of SSF system.

Lonsane and Ramesh reviewed the production of bacterial thermostable a -amylases by SSF, which they referred to as the potential tool for achieving economy in enzyme production and starch hydrolysis. Various methods to reduce the cost of production were discussed, taking into consideration enzyme production by *B. amyloliquefaciens* and *B. licheniformis*.

Numerous other microorganisms like *Saccharomycopsis capsularia* , *B. coagulans* , *Bacillus*sp. HOP-40 , and *B. megatarium* 16 M (ref. 187) have also been used for a -amylase production by SSF using agro-industrial residues.

Recovery of the enzymes from the fermented matter is an important factor that affects the cost-effectiveness of the overall process. In a significant finding, Padmanabhan *et al.* reported that the recovery of a -amylase from the solid fermented matter depended on the temperature of extraction. When enzyme was extracted and recovered at 50°C, the quantum of recovery was 2.2 fold higher than at 30°C. A further increase of about 19% in leaching efficiency was observed when contact time was extended from 60 to 120 min.

The other important enzyme of the amylase family is glucoamylase (GA). Traditionally, glucoamylase has been produced by SmF and one-way process in solution has been well developed. In recent years, however, the SSF processes have been increasingly applied for the production of this enzyme.

A strain of *A. niger* was used for the production of glucoamylase in solid cultures. The study included screening of a number of agro-industrial residues including wheat bran, rice bran, rice husk, gram flour, wheat flour, corn flour, tea waste, copra waste, etc., individually and in various combinations. Apart from the substrate's particle size, which showed profound impact on fungal growth and activity, substrate-moisture content and water activity also significantly influenced the enzyme's yield. Different types of bioreactors were used to evaluate their performances. These included flasks, aluminium trays, and glass columns (vertical and horizontal). Enzyme production in trays occurred optimally in 36 h in comparison to typically required 96 h in flasks. In a significant study on the effect of yeast extract on glucoamylase synthesis by *A. niger* NCIM 1248 in SSF system, it was observed that supplementation with 0.5% yeast extract resulted in about 20% increase in enzyme yields. GA was purified 32.4 fold with the final specific activity of 49.25 U/mg protein. Four different forms (GA-I, GA-I', GA-II, and GA-II'), having different characteristics were reported. This was the first report on the four forms of GA produced by *A. niger* by SSF.

There are reports describing a comparative profile of glucoamylase production in SmF and SSF systems. Interestingly, contrary to the general findings, Fujio and Morita reported a 4.6-fold lower glucoamylase yield by *Rhizopus* sp. A-11 in a conventional SSF process using wheat bran medium than by SmF which used metal-ion supplemented medium. Solid and liquid cultures yielded 150 and 189 mg of protein, respectively. Hata *et al.* compared the two glucoamylases produced in SmF and SSF systems using *A. oryzae*. Enzyme produced by SSF could digest raw starch but that by SmF could not. GA obtained by the two systems exhibited different characteristics. Tani *et al.* too compared characteristics of GA produced by either SmF and SSF processes. Solid culture was more efficient than liquid culture for GA production.

Rajgopalan *et al.* used a bacterial strain of *B. coagulans* for modelling of substrate-particle degradation in SSF system of GA. Enzyme diffusion was found to be a critical factor in degradation of the substrate particle. Mitchell *et al.* studied an empirical model of growth of *R. oligosporus* in SSF system. An equation was developed to describe glucoamylase activity on the substrate, which was then used to predict the growth. Apart from an early discrepancy, the growth rate correlated reasonably with the GA activity. Elegado and Fujio screened 39 *Rhizopus* isolates and 9 authentic *Rhizopus* strains (grown on wheat bran in a SSF system) for their soluble starch digestive GA (SSGA) and raw starch digestive GA (RSGA) activities. Results showed that these strains could be classified into four groups, based on their SSGA and RSGA production and ratio of SSGA to RSGA. Soccol *et al.* also screened 19 *Rhizopus* strains for their ability to grow on raw cassava. Only three strains grew significantly, and GA production was higher on raw cassava than on cooked cassava.

A patent was granted to Snow Brand Milk Prod in 1990 for a process for GA production on multi-stage culture medium. An effective method for GA production in SSF was also described by Kobayashi *et al.*. There are many other reports on GA production in SSF systems using different strains on various substrates.

Critical analysis of the literature shows that production of industrial enzymes by SSF offers several advantages. It has been well established that enzyme titres produced in SSF systems are many-fold more than in SmF systems. Although the reasons for this are not clear, this fact is kept in mind while developing novel bioreactors for enzyme production in SSF systems. It is hoped that enzyme production processes based on SSF systems will be the technologies of the future. Genetically improved strains, suitable for SSF processes, would play an important role in this.

4

Microorganisms in Fermentation Technology

Microorganisms are capable of growing on a wide range of substrates and can produce a remarkable spectrum of products. The relatively recent advent of in vitro genetic manipulation has extended the range of products that may be produced by microorganisms and has provided new methods for increasing the yields of existing ones. The commercial exploitation of the biochemical diversity of microorganisms has resulted in the development of the fermentation industry and the techniques of genetic manipulation have given this well-established industry the opportunity to develop new processes and to improve existing ones.

The term fermentation is derived from the Latin verb fervere, to boil, which describes the appearance of the action of yeast on extracts of fruit or malted grain during the production of alcoholic beverages. However, fermentation is interpreted differently by microbiologists and bio- chemists. To a microbiologist the word means any process for the production of a product by the mass culture of microorganisms. To a biochemist, however, the word means an energy-generating process in which organic compounds act as both electron donors and acceptors, that is, an anaerobic process where energy is produced without the participation of oxygen or other inorganic electron acceptors.

The process of fermentation is defined as "a biological process that occurs in the absence of oxygen or under anaerobic conditions. The word "Fermentation" originates from verb of Latin origin "fervere" which means to boil. However in the modern times the Industrial fermentation is used for large- scale cultivation of microorganisms which are largely aerobic in nature. Instead of fermentation technology, the term Bioprocess technology is more in use as bioprocessing involves a variety of enzyme-catalysed reactions carried out by living cells (or cell free systems) in industrial setups. The most important equipment in use for bioprocessing is a Fermenter or a Bioreactor.

Commercially Important Fermentation

1. Microbial cells or Biomass as the product: Eg. Bakers Yeast, Lactic acid bacillus, Bacillus sp.

2. Microbial Enzymes: Catalase, Amylase, Protease, Pectinase, Glucose isomerase, Cellulase, Hemicellulase, Lipase, Lactase, Streptokinase etc.
3. Microbial metabolites :
   - Primary metabolites – Ethanol, Citric acid, Glutamic acid, Lysine, Vitamins, Polysaccharides etc.
   - Secondary metabolites: All antibiotic fermentation
   - Recombinant products : Insulin, HBV, Interferon, GCSF, Streptokinase
   - Biotransformations: Eg. Phenyl acetyl carbinol,Steroid Biotransformation

Nutrient Sources for Industrial Fermentation

Medium for Industrial Fermentations

Any Microbe requires Water, Oxygen, Energy source, Carbon source, Nitrogen source and Micronutrients for the growth.

Carbon & Energy source + Nitrogen source + O2 + other requirements ® Biomass + Product + byproducts + CO2 + H2O + heat

Nutrient Sources for Industrial Fermentation

Nutrient Raw material

Carbon Source

Glucose Corn sugar, Starch, Cellulose

Sucrose Sugarcane, Sugar beet molasses

Lactose Milk whey

Fats Vegetable oils

Hydrocarbons Petroleum fractions

Nitrogen Source

Protein Soybean meal, Cornsteep liquor, Distillers' solubles

Ammonia Pure ammonia or ammonium salts

Nitrate Nitrate salts

Nitrogen Air

Phosphorous source Phosphate salts

Trace elements : Fe, Zn, Cu, Mn, Mo, Co

Antifoaming agents: Esters, Fatty acids, Silicones, Sulphonates, Polypropylene

Buffers: Calcium carbonate, Phosphates

Growth factors: Some microorganisms cannot synthesize the required cell components themselves and need to be supplemented: E.g. Thiamine, Biotin, Calcium pentothenate

Precursors: Directly incorporated into the desired product: Phenyl ethylamine into Benzyl penicillin, Phenyl acetic acid into Penicillin G

Inhibitors: To get the specific products: e.g. Sodium barbital for Rifamycin

Inducers: Majority of the enzymes are inducible and are synthesized in response of inducers: e.g. Starch for Amylases, Maltose for Pollulanase, Pectin for Pectinase

Chelators: Chelaters are the chemicals used to avoid the precipitation of metal ions. Chelaters like EDTA, Citric acid, Polyphosphates are used in low concentrations.

Biotechnology Companies Developing Pharmaceuticals via Fermentation in Transgenic Microbes or Mammalian Cell Cultures:

- Atlantic BioPharmaceuticals, Inc.
- Chiron Corporation
- BioPharmaceuticals Pipeline
- Genentech, Inc.
- Interomex Biopharmaceuticals
- Lonza Biotechnology
- Marathon Biopharmaceuticals, Inc.
- NPS Pharmaceuticals
- Protein Design Labs
- Protein Sciences Corp

PRODUCTS OF FERMENTATION PROCESSES

The growth of micro-organisms or other cells results in a wide range of products. Each culture operation has one or few set objectives. The process has to be monitored carefully and continuously, to maintain the precise conditions needed and recover optimum levels of products. Accordingly, fermentation processes aim at one or more of the following:

a. production of cells (biomass) such as yeasts;
b. extraction of metabolic products such amino acids, proteins (including enzymes), vitamins, alcohol, etc., for human and/or animal consumption or industrial use such as fertiliser production;
c. modification of compounds (through the mediation of elicitors or through biotransformation); and
d. production of recombinant products.

A. Microbial Biomass

Microbial biomass is produced commercially as **single cell protein**(SCP) using such unicellular algae as species of *Chlorella* or *Spirulina*for human or animal consumption, or viable yeast cells needed for the baking industry, which was also used as human feed at one time. Bacterial biomass is used as animal feed. The biomass of *Fusariumgraminearum* is also produced, for a similar use.

B. Microbial Metabolites

i) Primary metabolites:

During the log or exponential phase organisms produce a variety of substances that are essential for their growth, such as nucleotides, nucleic acids, amino acids, proteins, carbohydrates, lipids, etc., or by- products of energy yielding metabolism such as ethanol, acetone, butanol, etc. This phase is described as the tropophase, and the products are usually called primary metabolites.

ii) Secondary metabolites:

Organisms produce a number of products, other than the primary metabolites. The phase, during which products that have no obvious role in metabolism of the culture organisms are produced, is called the **idiophase**, and the products are called **secondary metabolites**.

In reality, the distinction between the primary and secondary metabolites is not a straightjacket situation. Many secondary metabolites are produced from intermediates and end products of secondary metabolism. Some like those of the Enterobacteriaceae do not undergo secondary metabolism.

Secondary metabolism may be **repressed** in certain cases. Glucose represses the production of actinomycin, penicillin, neomycin and streptomycin; phosphate represses streptomycin and tetracyclin production. Hence, the culture medium for secondary metabolite production should be carefully chosen.

C. Production Enzymes

Industrial production of enzymes is needed for the commercial production of food and beverages. Enzymes are also used in clinical or industrial analysis and now they are even added to washing powders (cellulase, protease, lipase). Enzymes may be produced by microbial, plant or animal cultures. Even plant and animal enzymes can be produced by microbial fermentation. While most enzymes are produced in the tropophase, some like the amylases (by *Bacillus stearothermophilus*) are produced in the idiophase, and hence are secondary metabolites.

D. Food Industry Products

A very wide range of innumerable products of the food industry, such as sour cream, yoghurt, cheeses, fermented meats, bread and other bakery products, alcoholic beverages, vinegar, fermented vegetables and pickles, etc., are produced through microbial fermentation processes. The efficiency of the strains of the organisms used, and the processes are being continuously improved to market quality products at more reasonable costs.

E. Recombinant Products

Recombinant DNA technology has made it possible to introduce genes from any organism into micro-organisms and vice versa, resulting in transgenic organisms and the latter are made to produce the gene product. Genetically manipulated Escherichia coli, Saccharomyces cerevisiae, other yeasts and even filamentous fungi are now being used to produce interferon, insulin, human serum albumin, and several other products.

F. Biotransformation

Production of a structurally similar compound from a particular one, during the fermentation process is transformation, orbiotransformation, or bioconversion. The oldest instance of this process is the production of acetic acid from ethanol.

Immobilised plant cells may be used for biotransformation. Using alginate as the immobilising polymer, digitoxin from Digitalis lanatawas converted into digoxin, which is a therapeutic agent in great demand. Similarly, codeinone was converted into codeine and tyrosine from Mucuna pruriens was converted into DOPA.

G. Elicitors

It is possible to induce production or enhance production of a compound in cultures by using **elicitors**, which may be micro-organisms. For example, *Saccharomyces cerevisiae* was an efficient elicitor in the production of glyceollin (Glycine max) and berberine (Thalictrum rugosum). Rhizopus arrhizus trebled diosgenin production by Dioscorea deltoidea. The production of morphine and codeine byPapaver somniferum was increased 18 times by Verticillium dahliae.

A BRIEF HISTORY OF FERMENTATION IN WEST

The origins of microbiology (other than the general knowledge of fermented foods which existed worldwide since ancient times) can be traced back to the invention of the compound microscope in the late 1500s. This relatively simple tool soon revolutionized man's knowledge of the heretofore invisible microbial world. In 1675 the Dutch merchant Anton van Leeuwenhoek, the greatest of the early microscopists, saw and reported one-celled organisms, which he called "animacules." (Today they are called "protozoa.") The discovery electrified the scientific world of the time. Then in 1680, using a microscope that magnified the diameter of each object 300-fold, he looked at yeast and found them to consist of tiny spheroids. While the protozoa were clearly alive, the yeast did not appear to be. No connection was drawn between the existence of these tiny organisms and the well known phenomenon of fermentation. So for 150 years after van Leeuwenhoek's pioneering observations, it was hardly thought that these minute organisms could be important enough to deserve serious study.

The early 1800s saw a great increase of interest in microbiology in Europe. The scientific period began with great advances in botany, increased interest in microscopy, and willingness to investigate individual organisms. The two major problems that would challenge the greatest researchers in the new field of microbiology concerned the basic nature of the fermentation process and the basic nature of enzymes. The scientific breakthroughs that would lead to the unraveling of the mysteries of fermentation starting in the 1830s were made primarily by French and German chemists.

In the late 1700s Lavoisier showed that in the process of transforming sugar to alcohol and carbon dioxide (as in wine), the weight of the former that was consumed in the process equaled the weight of the latter produced. In 1810 J.L. Guy-Lussac summarized the process with the famous equation CÚ6ØHÚ12ØOÚ6Ø yields 2COÚ2Ø + 2 CÚ2ØHÚ6ØO. The entire process was considered to be simply a chemical reaction and yeast (which was not yet even classified as a definite substance, much less a living organism instrumental to fermentation) was thought to play a physical rather than a chemical role, an idea dating back to the time of Georg Stahl in 1697. It was held that either the catalytic action at the yeast cell or the molecular vibrations from the decomposing organic matter arising from the death of the cells, sparked the chemical changes resulting in fermentation. Putrefaction, spoilage, and fermentation were all considered to be processes of death, not life.

The first solid evidence of the living nature of yeast appeared between 1837 and 1838 when three publications appeared by C. Cagniard de la Tour, T. Swann, and F. Kuetzing, each of whom independently concluded as a result of microscopic investigations that yeast was a living organism that reproduced by budding. The word "yeast," it should be noted, traces its origins back to the Sanskrit word meaning "boiling." It was perhaps because wine, beer, and bread were each basic foods in Europe,that most of the early studies on fermentation were done on yeasts, with which they were made. Soon bacteria were also discovered; the term was first used in English in the late 1840s, but it did not come into general use until the 1870s, and then largely in connection with the new germ theory of disease.

The view that fermentation was a process initiated by living organisms soon aroused fierce criticism from the finest chemists of the day, especially Justus von Liebig, J.J. Berzelius, and Friedrich Woehler. This view seemed to give new life to the waning mystical philosophy of vitalism, which they had worked so hard to defeat. Proponents of vitalism held that the functions of living organisms were due to a vital principal (life force, *chi, ki, prana* , etc.) distinct from physico-chemical forces, that the processes of life were not explicable by the laws of physics and chemistry alone, and that life was in some part self determining. As we shall soon see, the vitalists played a key role in debate on the nature of fermentation. A long battle ensued, and while it was gradually recognized that yeast was a living organism, its exact function in fermentations remained a matter of controversy. The chemists still maintained that fermentation was due to catalytic action or molecular vibrations.

The debate was finally brought to an end by the great French chemist Louis Pasteur (1822-1895) who, during the 1850s and 1860s, in a series of classic investigations, proved conclusively that fermentation was initiated by living organisms. In 1857 Pasteur showed that lactic acid fermentation is caused by living organisms. In 1860 he demonstrated that bacteria cause souring in milk, a process formerly thought to be merely a chemical change, and his work in identifying the role of microorganisms in food spoilage led to the process of pasteurization. In 1877, working to improve the French brewing industry, Pasteur published his famous paper on fermentation, *Etudes sur la Biere* , which was translated into English in 1879 as *Studies on Fermentation* . He defined fermentation (incorrectly) as "Life without air," but correctly showed specific types of microorganisms cause specific types of fermentations and specific end products. In 1877 the era of modern medical bacteriology began when Koch (a German physician; 1843-1910) and Pasteur showed that the anthrax bacillus caused the infectious disease anthrax. This epic discovery led in 1880 to Pasteur's general germ theory of infectious disease, which postulated for the first time that each such disease was caused by a specific microorganism. Koch also made the very significant discovery of a method for isolating microorganisms in pure culture.

Interestingly, until his death in 1873, the eminent German chemist J. von Liebig continued to attack Pasteur's work on fermentation, putrefaction, and infectious diseases. He recognized the similarity of these phenomena but refused to believe that living organisms were the main causative agents. Fermentation, he felt, was primarily a chemical rather than a biological process. History has shown, with the discovery of enzymes, that Pasteur was not entirely right, nor Liebig entirely wrong.

The work of Pasteur and his many colleagues and predecessors opened up vast new vistas in the fields of biochemistry, microbiology, and fermentation. The term "biochemistry" was first used in English in 1869, but this new science of the application of chemistry to biology was generally called "physiological chemistry" until the early 1900s. The two outstanding pioneers were Liebig and Pasteur. The term "microbiology" was first used in English in 1885, long after Pasteur's major discoveries. But basic knowledge of this new science of the study of minute living organisms closely related to human activity or welfare did not begin to enter the popular consciousness until the early 1900s. At about this time the scientific breakthroughs of the 1870s and 1880s had begun to produce a change in people's conception of the world around them so sweeping and profound as to be termed revolutionary. Food microbiology was finally set on a scientific foundation, based on the action of specific microorganisms. A rational theory of infectious diseases (which formerly were not differentiated from one another) set people's minds free from the age-old fear of vengeance from an unknowable and invisible disease-causing entity. And the ancient theory of spontaneous generation of lower life forms, which said they could arise *de novo* and fully formed from decomposing matter, was replaced by the verifiable theory of biogenesis. For the first time people began to accept the fact that they shared their environment with multitudes of minute organisms that exerted an ongoing powerful influence on human life. This new world view, among other things, provided a tremendous stimulus for new research on fermented foods.

Although showing that fermentation was generally the result of the action of living microorganisms was an epic breakthrough, it did not explain the basic nature of the fermentation process, or prove that it was caused by the microorganisms that were apparently always present. As early as the late 1700s it had been recognized that there was another type of chemical change that resembled the yeast fermentation in some respects. This was the sort of changes that occur, for example, in the digestion of food. In 1752 Reamur, in studying the digestive processes of a falcon, showed that its digestive juices were able to dissolve meat. In 1785 William Irvine discovered that aqueous extracts of sprouted barley caused liquefaction of starch. The first clear recognition of what were later called "enzymes" came in 1833 when two French chemists, A. Payen and J.F. Persoz, made a more detailed investigation of the process of solubilizing starch with a malt extract to form a sugar that they called "maltose." They called the agent responsible for this transformation "diastase" and they showed that it was destroyed or inactivated by boiling, that without undergoing permanent change itself, a small amount of diastase could convert a large amount of starch to sugar, and that it could be concentrated and purified by precipitation with alcohol. In 1835 the German naturalist Swann, mentioned above for his early work with fermentation, isolated a substance from gastric juice which could bring about the dissolution of meat but which was not an acid. He called it "pepsin" from a Greek word meaning "digestion." It soon became fashionable to call organic catalysts such as diastase and pepsin "ferments," because digestion and fermentation, both allied with life, seemed to be somewhat similar processes. Under the influence of the vitalists, ferments were grouped into two types: those involved with life process were called "organized ferments" and those which were not (like pepsin) were merely "unorganized ferments." A relation between the two types of ferments was suspected by many, and in 1858 M. Traube put forward the theory that all fermentations were due to ferments, definite chemical substances he regarded as related to the proteins and produced in the cells by the organism. In 1876, to reduce confusion that existed concerning the two types of ferments, the German physiologist Wilhelm Kuehne suggested that an unorganized ferment, acting in the absence of life, be called an "enzym," after the Greek words meaning "in yeast;' in 1881 this term was anglicized to "enzyme" by William Roberts, and it had begun to catch on by the 1890s.

Many scientists, including Pasteur, had attempted unsuccessfully to extract the fermentation enzyme from yeast. Success came finally in 1897 when the German chemist Eduard Buechner ground up yeast, extracted a juice from them, then found to his amazement that this "dead" liquid would ferment a sugar solution, forming carbon dioxide and alcohol... just like living yeasts. Clearly the so-called "unorganized ferments" behaved just the organized ones. From that time on the term "enzyme" came to be applied to all ferments. The term "ferment" dropped out of the scientific vocabulary altogether and the vitalist position collapsed, never to recover. Thereafter it was agreed that only one set of laws applied to all things, both animate and inanimate, and that there was no special vital force which characterized living things and acted under different laws. And it was finally understood that fermentation is caused by enzymes which are produced by microorganisms. In 1907 Buechner won the Nobel Prize in chemistry for his work, which opened a new era in enzyme and fermentation studies.

The sciences of microbiology, biochemistry, fermentation technology, mycology, and bacteriology all shared a deep interest in the nature and working of enzymes.

Yet still by the early 1900s no one knew exactly what enzymes were or how they acted. As the agricultural microbiologist Conn asked in 1901, "How can they produce chemical actions without being acted upon or entering into the reactions? Are enzymes fully lifeless or semi-living? We still do not know the fundamental mystery of fermentation." Gradually an understanding of enzymes and catalysts developed. In 1905 Harden and Young discovered coenzymes, agents necessary for the action of enzymes. In 1926 the American biochemist J.B. Sumner first purified and crystallized an enzyme (urease) and showed that it was a protein, more precisely a protein catalyst. Eventually enzymes came to be seen as the key catalysts in all the life processes, each highly specialized in its catalytic action and generally responsible for only one small step in complex, multi-step biochemical reactions. Enzymes are still produced only by living organisms, both animals and plants; they have never?? been synthesized.

Advances in microbiology and fermentation technology have continued steadily up until the present. For example, in the late 1930s it was discovered that microorganisms could be mutated with physical and chemical treatments to be higher yielding, faster growing, tolerant of less oxygen, and able to use a more concentrated medium. Strain selection and hybridization developed as well, affecting most modern food fermentations (Hesseltine and Wang 1977).

A Brief History of Fermentation in East Asia. Traditional fermented foods play an unusually extensive role in East Asia food systems. These fermented foods have a number of important distinguishing characteristics: a number of the most important fermentations use molds; dairy products and other animal proteins (excepting fish) are not widely used, as they are in the West; and modern fermentation processes and technology are based largely on traditional processes, yet are extremely advanced and sophisticated.

The main use of molds has been in the process of making koji (mold-fermented grains and/or soybeans), which serves as a source of more than 50 enzymes in a subsequent fermentation in much the same way that, in the West, the enzymes of malt (steeped and sprouted barley or other cereal grains) are used to make alcoholic beverages.

The nature of koji is embodied in the very characters with which the word is written. In the more traditional form—used with most miso koji and especially with barley koji—the ideographs for "barley:" and "chrysanthemum" are placed side by side. In the more recent form—used especially with ready-made rice koji—the ideographs for "rice" and "flower" are conjoined. The first form is said to have originated in China, whereas the latter was developed in Japan about 1,000 years ago. In both, the notion of grain covered with a bloom of mold is vividly expressed (Shurtleff and Aoyagi 1976; Tamiya 1958).

Since ancient times the koji making process has been unique to East Asia, where it has been used in the preparation of fermented foods such as miso, soy sauce, soy nuggets, sake, shochu (spirits), and rice vinegar (*yonezu*). The only traditional East Asian fermented soyfood not prepared with molds is Japan's natto, and its relatives *thua-nao* in Thailand and *kinema* in Nepal; these are bacterial fermentations. Some have suggested that molds are widely used since they grow well in areas having a humid climate and long rainy season during the warm months. In the West mold fermented foods are limited primarily to a number of cheeses characterized by their strong flavors and aromas: Camembert, Blue, Brie, and related types. Because of the widespread use of mold-fermented foods in East Asia, the word "mold" there has a rather positive connotation, something like "yeast" in the West. Most Westerners still have a deep-seated prejudice against moldy products, and they generally associate the word "mold" with food spoilage, as in "moldy bread."

Surprisingly little has been published in English about the history of fermentation and knowledge of the fermentation process in East Asia, especially the history prior to the 1870s and 1880s, when the new science of microbiology was introduced from the West. The few works that do exist will be cited later.

The earliest records of the koji-making process can be traced back to at least 300 BC in China and to the third century AD in Japan. Molds differ in one important respect from yeasts and bacteria in that they can be easily observed with the naked eye (without a microscope) and their growth, form, and color noted. In East Asia it was probably understood that fermentation was a life process long before it was in the West. By the sixth century AD, as recorded in the *Ch'i-min yao-shu* (the earliest encyclopedia of agriculture), the Chinese had distinct names for two types of molds used in fermented soyfoods; what we now call *Aspergillus* was then called "yellow robe" and *Rhizopus* was called "white robe." These cultures were carefully distinguished and propagated from year to year. By the 10th century a koji starter or inoculum was deliberately being used in the preparation of koji for fermented foods (Tamiya 1958; Sakaguchi 1972; 1979).

From these early times until the 1870s the traditional fermented foods industries in East Asia apparently advanced largely by an empirical, trial-and-error process without the benefit of general scientific research into the nature of microorganisms and of the fermentation process, and without any general theories in these areas.

From 1635 until 1854, Japan lived in relative seclusion and isolation under the Tokugawa shoguns. The great advances made in Western science during this 220 year period passed largely unnoticed. Then in 1854 the American Commodore Perry and two fully-armed steamships arrived in Japan and demanded that the country end its self-imposed isolation and open itself to trade with the West. Internal pressures and the intrusion of the Western powers helped topple the already declining Tokugawa Shogunate in 1868, and that year the emperor was restored as the political head of the nation, ushering in the Meiji Restoration. Openness, modernization, westernization, scientism, positivism, and the ideal of progress all formed the dominant ethos of the Meiji Period, which lasted until 1912. This new spirit corresponded with a golden age in the West of scientific and technological breakthroughs and empire building, which gave Westerners confidence in their ability to control the world and an optimistic faith in a bright new future. In Japan, Western knowledge, science, and technology were actively sought and cultivated, for it was generally thought that the very survival of the nation depended on their quick assimilation. By the 1870s governmental colleges had been established primarily for teaching Western science and technology. By paying princely salaries and offering high positions, the Japanese were able to attract top European scientists to staff the new universities, the foremost of which was Tokyo Imperial University, established in 1877. At the same time, for the first time in 220 years, Japanese were allowed to travel abroad; scientists and students were sent abroad for training. As Atkinson (1881b), one of the earlier exchange scientists to Japan, noted:

The student of science in Japan has a wide field before him; that system of isolation which has prevented the introduction of Western knowledge till within the last quarter century has not been entirely fruitless, for it has resulted in the development of industrial processes which are as novel and interesting to the European as those of the latter are to the Japanese.

The imported European scientists and professors caused an almost immediate revolution in the field of East Asian food fermentations, for they brought both the powerful tools of the Western scientific method and a host of new discoveries in the fields of fermentation and microbiology. In Japan the effect of German (and to a lesser extent English) microbiologists and chemists was initially most pronounced. Some European countries sent their scientists out to their colonies; the Dutch (such as Went and Prinsen Geerligs) went to the Netherlands East Indies (Indonesia) and the French went to Indochina. In each area fermented soyfoods were investigated. It is interesting to note that no such studies of fermented foods were done in China by either Western or Chinese researchers during this vital pioneering period, for various reasons: China was slow to modernize and Westernize; the late 1900s were a period of rebellion and decline; and Western imperialism, so destructive to China, had made the Chinese closed to and suspicious of Western ideas. Nevertheless China, recognizing the rapid advance of Japanese science from its interaction from the West, eventually chose to get its information from the West indirectly via Japan (Hesseltine and Wang 1977).

Prior to 1870, makers of East Asian fermented foods were unaware of the basic nature of the fermentation process of microorganisms, enzymes, and their respective interactions. Makers of koji had no idea what caused the grains and/or soybeans to become covered with a fragrant white mycelium after several days of incubation in a warm koji room, or what later transformed the koji almost magically into delicious, savory seasonings such as miso, shoyu, or soy nuggets, or into heady beverages such as sake. The microscope was essentially unknown in East Asia prior to the 1880s.

The advances in food fermentations resulting from the exchange of people and ideas was most pronounced in Japan. The first generation of European scientists there plunged in to their investigations of the many fermented foods with great curiosity and enthusiasm. One of their first subjects of research was the koji mold, now known as *Aspergillus oryzae* (pronounced ass-per-JIL-us oh-RAI-zee), and the various foods in which it was used, especially sake and shoyu, which were major sources of tax revenue for the Meiji government. Tradition ascribed the introduction of sake brewing in Japan to some emigrants from Korea at about the end of the third century AD; they doubtless learned the process from China, where it had long been practiced. One of the earliest accounts of sake production by a Westerner appeared in 1874 when Dr. J.J. Hoffmann, a German professor in the medical school of today's Tokyo University, published a translation of an article on sake from a Japanese encyclopedia of 1714. The same year he wrote a detailed and scientific description of the process for making rice koji, based on his visits to the famous plant run by Mr. Sagamia Monjiro, which made sake, mirin, and shoyu in Nagareyama, 5 miles north of Edo (Tokyo). Although Hoffmann described accurately the process for making rice koji, he did not use the word "koji." In 1878 Korschelt, also a German, gave an extremely accurate and detailed (1600-word) description of exactly how koji and koji starter were made. He was the first Western scientist to use the words *koji* and *tane koji* (koji starter); he used them frequently and accurately. Actually the term *koji*appeared slightly earlier in Hepburn's famous Japanese-English Dictionary,which translated it (inaccurately) as "barm or yeast." In a section of Korschelt's article, and in a Japanese article written with Matsubara in 1878, Ahlburg, another German, who taught natural history at Tokyo University, gave the first detailed description of the koji mold, which he called *Eurotium Oryzae* Ahlburg. In 1884 (Saito 1906 says 1883), Ferdinand J. Cohn, a Polish botanist and microbiologist, first gave the koji mold its present name,*Aspergillus oryzae* . (The genus *Aspergillus* was first identified and named by Micheli in 1729. After 1884 the koji mold was referred to as *Aspergillus oryzae* (Ahlburg) Cohn, in recognition of Ahlburg's earliest accurate description. The mold's characteristics were subsequently clarified and elaborated by Buesgen (1885), Schroeter (1893, and especially Wehmer (1895, 1901).

Another pioneer in the field of koji research was Atkinson, who had a BS degree from London and was a professor of analytical and applied chemistry at Tokyo University. In 1878, after visiting sake factories, he wrote "On Sake Brewing," which contained a preliminary description of the koji-making process and mentioned the word "koji." In 1881, after extensive research with his assistant Mr. Nakazawa at the koji plant of Mr. J. Kameyama in Yushima near Tokyo, he published two major articles. In his 73-page "On the Chemistry of Sake Brewing," he gave a detailed account of koji making in underground caves in Tokyo and an analysis of its composition. His "On the Diastase of Koji" first demonstrated that the koji mold had strong diastatic (amylolytic) activity. In 1889 Dr. Oscar Kellner (a German Professor of Agricultural Chemistry at Tokyo University) and his Japanese co-workers published pioneering studies on koji, shoyu, and miso. Then in 1895 and 1901, C. Wehmer, who taught mycology at Hannover, described the koji mold in great detail. He also stated (1895) that koji was being made in America at a large Japanese sake brewery on U

Street in Peoria, Illinois, the very area that would become America's leading center of research on koji and miso, starting in the 1960s! As Western researchers studied koji, they quickly realized that it has much the same relationship to shoyu and miso fermentation that malt has to Western alcoholic grain fermentations.

The rich interchange between Japan and Europe, and between scientists and food manufacturers led to major benefits to all parties. As Atkinson (1881b) noted:

We cannot omit to here draw attention to the mutual advantage to be derived from an association of workers in industrial and in pure science; the cooperation cannot but be of the greatest utility on the one hand, by suggesting new subjects for research to the theoretical worker, and on the other, in aiding the practical man to attain the best results possible.

Most of Japan's imported European professors had bright Japanese students and technicians, whom they taught and trained carefully. By the late 1800s and early 1900s these students, and others who had gone to Europe to study, were publishing scientific articles in both Japanese and European journals about traditional fermented soyfoods and koji.

One of the first Japanese to make an important commercial application of the new knowledge of microbiology and fermentation science was Jokichi Takamine, who had studied at a Japanese university. Having heard of the malting process in the West, he decided ambitiously to try to introduce the koji process in its place. After increasing considerably the diastatic activity of the koji mold, he went to America in 1880, but met with opposition from the malt makers. He then undertook a new project to extract the enzymes from the koji mold for commercial use. In 1894 he was granted two US patents (Nos. 525,823 and 525,825) for a process for making diastatic enzyme, which was very successfully marketed as Takadiastase. This product, which contained a rich variety of enzymes, came to be used widely in the field of enzymology; it brought international fame to both him and the koji mold. Takamine was far ahead of his time in recognizing the potential industrial significance of enzymes, even in an era when knowledge of enzymes was very scanty. He later discovered the adrenal hormone, adrenaline/ epinephrine (Tamiya 1958).

Another early leader in the fields of microbiology and fermented soyfoods was K. Saito. He did excellent early investigations on the shoyu fermentation, named the primary tempeh mold (*Rhizopus oligosporus*) in 1905, and was an authority on yeasts and molds. Likewise K.N. Yabe did important early work in bacteriology and in natto fermentation.

Two other early pioneers in the introduction of microbiology and fermentation science to Japan were Dr. Teizo Takahashi (1875-1952) and his brilliant student Dr. Kinichiro Sakaguchi (1897-), both of whom were professors in the Department of Agricultural Chemistry of Tokyo University. An excellent book chronicling the contributions of these two men and containing summaries of 190 papers relating to Dr. Takahashi's work and 390 relating to Dr. Sakaguchi's has been published by Asai and Arima (1958) and a Commemorative Committee. Both men did numerous important studies relating to miso, shoyu, and the koji mold, *Aspergillus* (see Bibliography). In addition, Dr. Sakaguchi was deeply interested in the history of fermentation and fermented foods (including fermented soyfoods) in East Asia, and he wrote some of the best works in English available on this subject (Sakaguchi 1972, 1979).

Hesseltine and Wang (1977) noted eight areas in which Western microbiologists have made contributions to indigenous fermented foods. In addition to (1) training teachers and technicians, and (2) studying all scientific aspects of the fermentation process, they have (3) introduced breeding of microorganisms for strain improvement (as with shoyu and miso), (4) promoted use of pure cultures in the fermentation (of all fermented soyfoods), (5) described the changes in the substrate during fermentation (especially with tempeh and miso), (6) established the food values of the products, (7) suggested new food uses of the products (especially tempeh and miso), and (8) developed an awareness of the importance of studying indigenous fermented foods. A final contribution might be the development of new technologies transferable to traditional fermented foods; an example would be the perforated polyethylene bags for tempeh incubation developed at the USDA Northern Regional Research Center.

During the 20th century, Japanese microbiologists have made many important contributions to the development of applied and industrial microbiology, including the manufacture of fermented soyfoods, as well summarized by Tamiya (1958) and Sakaguchi (1972). Until quite recently, their strength was more in the area of application of scientific knowledge than in pioneering basic scientific and microbiological breakthroughs. From the early 1900s, important studies on the koji mold and its enzymes were done by Japanese scientists. Important advances in enzymology, with much of the work done on koji molds, began in the 1920s. In 1928 Miyazaki developed the combined Amylo-Koji process. By the 1950s Japanese scientists had isolated various protease and amylase enzymes, induced mutations, and used them commercially. They also developed the technology for the microbial production of L-glutamic acid and monosodium glutamate (MSG), lysine and other amino acids, flavor enhancing nucleotides such as inosinic acid, and organic acids. They used the koji mold *Aspergillus oryzae* in the commercial production of enzymes including proteases, amylases, amyloglucosidase, and lipase. They made microbial rennet and numerous other products. Indeed in the period following World War II, Japan became the world leader in the field of industrial fermentations. Wang and Hesseltine (1979) have suggested that this may have been “in large part due to the food fermentation base from which it launched its industrialization of micoorganisms.”

According to Tamiya, in 1958 food and drinks produced with koji retailed for $1,000 million a year, and the taxes from these foods amounted to more than $500 million, which was as much as 20% of the Japanese national budget! In 1970 in Japan, foods made from koji molds accounted for 1.5% of the nation’s Gross National Product, or about $205,000 million. Prominent among these were miso and shoyu (Sakaguchi 1972). Production of fermented soyfoods continues to be the most important of the fermented food industries of East Asia.

Starting in about the 1960s and increasing rapidly after the mid-1970s, East Asian fermented soyfoods (especially soy sauce or shoyu, miso, and tempeh, in that order), began to be widely used in the West. Reasons for this include the growing general interest in soyfoods, the cultural and religious movement toward meatless and vegetarian diets, the increasing interest in nutritious foods with less animal fats, the awareness these foods as a good vegetarian source of vitamin B-12, the growing worldwide travel stimulating interest in foreign foods, the increase of East Asian refugees to the West, and the increased interest in microbiology and enhanced image of fermented foods. All of these developments indicate a bright future for fermented soyfoods in the West (Hesseltine 1981).

Role of Microorganisms in The Fermentation Process

Initially, it might be hard to imagine how microorganisms contribute to the fermentation process. How can such small and microscopic cells unseen to the naked eyes produce such high volumes of products? How can such microorganisms produced such a rich diversity of fermentation products which are useful to human beings?

The answer to the first question is simple enough. One tiny microscopic microbial cell will not produce industrially significant volume of fermentation products. However, when we talk about billions and billions of microbial cells in the fermentation vat, the volume and concentration of fermentation products is industrially and economically significant!

The diversity of fermentation products produced by the microorganisms is attributed to the rich diversity of microorganisms which have a diverse metabolisms that can yield various types of fermentation products.

Why Microorganisms are the Best Choice For Fermentation?

The success of using microorganisms for fermentation lies in their very microscopic and metabolic characteristics. It is good being small!

High Surface area to Volume Ratio

Microorganisms are very very tiny creatures. Taking an example of a rod bacterium (imagine a brick!), we can see that it has six free surfaces that surrounds the bacterium. These six free surfaces interfaced with the surrounding environment from where they obtained their nutrients or to where they throw away their metabolic waste products.

With such a high number of free surface areas in a tiny volume of cell, it confers upon the bacterial cell a very high surface area to volume ratio. This very high surface area to volume ratio allows maximum or optimum surfaces for diffusions or molecular exchanges to occur between the microbial cell and the environment. No matter where the molecules are, they are easily accessible for diffusion into the microbial cell.

Compare this with the elephant which is such a huge animal. which has a very low ratio of surface area to volume ratio! If nutrients are to diffuse through the surface area of the elephant it would not be enough to supply every cell of the elephant!

Once the nutrient molecules diffuse through the cell wall and membrane, it can be easily transported to where its needed in the cell. The size of the cell is so microscopic that distance covered in the transportation of the molecules in the cell is a convenience!

With such an ease of diffusion of nutrient molecules from the environment into the cell, and the diffusion of waste products from within the cell outwards to the surrounding environment we can see that metabolism will be at optimum state.

The efficient nutrient uptake coupled with the small size of the cell will allow for rapid synthesis and reproduction of new cells. Microorganisms under ideal state will double up within hours. Animals like elephants and human may take months to reproduce themselves.

Mode of Nutrients Transportation

The nutrients which diffuse into the microbial can either use simple diffusion process which is powered by the differences in the concentration gradients between the environment and within the cell. For very small nutrient molecules, most would diffuse by the mechanism of passive diffusion. Larger and complex molecules use active or group transport which requires expansion or utilization of energy

Microbes easily reproduce asexually! There is no real need to have a opposite partner cells, get married and reproduce. They will just as easily split their cells into two daughter cells which will later grow into larger cells and repeat the cycle...

Genetic Adaptability

Microorganisms generally show the ability to adapt to new environment. They can get easily adapted to living under different environmental conditions and also adapting to new sources of carbon or substrate. This ability is the result of various genetic adaptation which selects "successful" strains through mutation and genetic recombination. Some of the bacteria are even equipped with plasmids which can synthesize new enzymes that help the microorganisms exploit the new environment. The very short generation times and the high population generation will aid the selection and recombination process.

Metabolic Diversity

The unique thing about microbes are their metabolic diversity shown by various members of the microorganisms. They have the ability to use different energy sources and to use different types terminal electron acceptors.

Their ability to use different substrates is also correlated with the microbes ability to produce a diversity of fermentation products.

Fermentation bacteria are anaerobic, but use organic molecules as their final electron acceptor to produce fermentation end-products. *S*treptococcus, Lactobacillus, and Bacillus, for example, produce lactic acid, while Escherichia and Salmonella produce ethanol, lactic acid, succinic acid, acetic acid, CO2, and H2.

Fermenting bacteria have characteristic sugar fermentation patterns, i.e., they can metabolize some sugars but not others. For example, Neisseria meningitidis ferments glucose and maltose, but not sucrose and lactose, while Neisseria gonorrhoeaferments glucose, but not maltose, sucrose or lactose. Such fermentation patterns can be used to identify and classify bacteria.

During the 1860s, the French microbiologist Louis Pasteur studied fermenting bacteria. He demonstrated that fermenting bacteria could contaminate wine and beer during manufacturing, turning the alcohol produced by **yeast** into acetic acid (vinegar). Pasteur also showed that heating the beer and wine to kill the bacteria preserved the flavor of these beverages. The process of heating, now called pasteurization in his honor, is still used to kill bacteria in some alcoholic beverages, as well as milk.

Pasteur described the spoilage by bacteria of alcohol during fermentation as being a "disease" of wine and beer. His work was thus vital to the later idea that human diseases could also be caused by microorganisms, and that heating can destroy them.

TRADITIONAL FERMENTATION TECHNOLOGY

Fermentation technology, as mentioned in the literary texts, is more than 3000 year old in India. The fermentation technology employed a variety of processes and was put to a large number of uses. It also laid the foundation of alchemy and chemistry.

The term fermentation is derived from the Latin word *Fermentum* that stands for boiling. Fermentation is the process of digesting certain substances that leads to chemical conversion of organic substances into simpler compounds. Fermentation technology began with sweet substances (vegetable or animal) in different parts of the world. The process of fermentation was probably discovered by observing the changes in the juices of several fruits and other substances that had been kept for a day or more. It appears that fermentation technology started simultaneously with settled agriculture during the Neolithic period.

A sweet substance known as *Soma* juice prepared by the Vedic Aryans is supposed to be the first product of fermentation in India. The *Rgveda* (c.1500 BC) shows that fermentation technology took its first step in connection with the preparation of *Soma* juice in India. There is also another drink, known as *Sura* (wine/beer), prepared by fermentation. These two preparations have also been used in different medicinal preparations, surgical procedures and in many chemical and alchemical operations. It is believed that acetic fermentation was known to India since the early times.

Curd is another very popular fermentation product described in the *Rgveda*. It still remains a popular food. The technology of curdling milk is also found in a number of texts associated with *Yajurveda*. In the beginning, fermentation was mainly associated with the preparation of spiritual drinks, but later on it was used for other purposes also. The different terms associated with fermentation in various texts are as follows:

- Asuta (begetting a new form)
- Abhisuta (extraction, may indicate extraction from fermented smash)
- Parisrut ('foaming', 'fermenting', that is, the state of fermenting)
- Sandhana (complete absorption of ferment with fermenting material)

Fermentation Generates Flavor in Bread

As Harold McGee, the author of *On Food & Cooking*, has pointed out, big molecules in proteins, starches, and fats don't have much flavor, but when they break down into their building blocks—proteins into amino acids, starches into sugars, or fats into free fatty acids—they all have marvelous flavors. Fermentation, whether it's acting on fruit juices to make wine or on flour to make bread, does exactly that—it breaks down large molecules into smaller, flavorful ones.

At the beginning of fermentation, enzymes in the yeast start breaking down starch into more flavorful sugars. The yeast uses these sugars, as well as sugars already present in the dough, and produces not only carbon dioxide and alcohol but also a host of flavorful byproducts such as organic acids and amino acids. A multitude of enzymes encourages all kinds of reactions that break big chains of molecules into smaller ones—amylose and maltose into glucose, proteins into amino acids.

As fermentation proceeds, the dough becomes more acidic. This is due in part to rising levels of carbon dioxide, but there are also more flavorful organic acids like acetic acid (vinegar) and lactic acid being formed from the alcohol in the dough. (This is similar to what happens to a bottle of wine that has been left uncorked for a while: the alcohol combines with oxygen to make vinegar.) The acidity of the dough causes more molecules to break down. The dough becomes a veritable ferment of reactions. Eventually, the amount of alcohol formed starts to inhibit the yeast's activity.

Yeast has help in producing flavorful compounds. Bacteria are important flavor builders as well. There are bacteria in the dough from the beginning, but as long as the yeast is very active, it consumes sugars as quickly as they're produced, leaving no food for the bacteria, which also like sugar. But when bakers chill a dough and slow down its rise, the cold dramatically reduces yeast activity. The bacteria, on the other hand, function well even in cold temperatures, so they now have an opportunity to thrive, producing many more marvelously flavorful acids.

Microbial Starter Cultures in Fermentation

The success of many food and beverage based fermentation industries depends heavily on the availability of the microorganisms which drive the fermentation process. The type or types of microorganisms used, the quality and volume of inocula used in the fermentation process is crucial in the fermentation industries. In reality, most of the food or beverage based fermentation industries use pure or mix cultures for their fermentation depending on the type of fermentation carried out. Some fermentation requires the use of mono cultures while there are food or beverage fermentations that require the contribution of a number of species of microorganisms for the fermentation to be successful.

The microorganisms used in the various food and beverage fermentations were in fact historically isolated from the ' natural contaminants' which occur on the decomposing or naturally fermented foods!

The use of starter cultures is not new. In traditional fermentations traditional starter cultures have always been used in the fermentation though many do not realize it. Some of these starter cultures used are obtained from:

1. Remnants or residuals of past fermented food and beverage
2. Original flora that existed naturally on wrappings used in the fermentation such as banana leaves
3. Inocula that occur naturally in the substrate of the fermentation such as grapes used in wine making
4. Traditionally made 'kojis' used in many soy sauce or tuak fermentation.

It is only now in the modern, sanitary, stainless steel fermentors of modern fermentation industries are ' starter cultures' used to start the desired fermentation process. Starter cultures are specially isolated and selected strains of microorganisms used in the fermentation which are massively cultivated under strict and scientific control to give the best and optimized fermentation which are not often seen in traditional food fermentations

Before the first cultures entered the market, traditional fermented food producers tended to rely on their own cultures. However, inadequately characterised cultures and inadequately defined compositions generally variability in the quality of the fermented products which is regarded not satisfactory in the market.

The development of such cultures requires the expertise and experience of specialists microbiologists, biochemists and food technologists as not all combinations are suitable for all products. The use of different strains in the same microbial starters must not result negative influence on each other. Usually in any microbial starters it is not wise mix more than four different species in a culture,"

The advantages of using industrially prepared starter cultures for food and beverage fermentation are:

1. Use only the microorganisms responsible for the fermentation
2 No other microbial contaminants that can cause diversion from the optimal fermentation
3 Can use the best strains specifically designed for the particular fermentation
4 Use of large inocula of starter cultures will result in better and faster fermentation
5 Quality and stability of the fermentation will be maintained
6 Taste, texture, colour stability and shelf life are mainly determined by microorganisms involved in the maturation, i.e. fermentation, of the product.

Fermentation technology is one of the oldest technologies employed in the food processing industry.

The term fermentation is derived from the Latin verb 'fervere', to boil, which describes the appearance of the action of yeast on extracts of fruits or malted grain during the production of alcoholic beverages.

In general it can be defined as the process of growing a culture of microorganisms in a nutrient medium at maintained physic-chemical conditions and thereby converting feed into desired end product.

Fermentation involves the breaking down of complex organic substance into smaller ones.

The microbial or animal cell obtains energy through glycolysis, splitting a sugar molecule and removing electrons from the molecules.

Fermented products encompass, but are not limited to wine, beer, vinegar, bread, soy sauce, sauerkraut, kimchi, pickled, olive, different fermented milk products, a large number of cheeses and a variety of sausages.

Many species of microorganisms are used for carrying out the process of fermentation to produce useful products. They include bacteria, fungi, algae and actinomycetes.

Several advantage of fermentation technology included:

- Produces value added and add variety to the human being's diets.
- Preservation of the food.
- Food quality improvement through flavor development, nutrient enrichments.
- The food more nutritious.
- Detoxifies of foods

Today large numbers of chemicals are produced by fermentation technology with the advent genetic engineering and the developments in computer technology.

Microorganisms and Fermentation Technology of Highland Barley Wine

Mold, yeast and lactic acid bacteria about Highland barley wine fermentation have been separated and identified on a large scale. By screening these microorganisms, two yeast strains ZY2 and ZY28, one mold strain ZM2 and one lactic acid bacteria strain ZB12 were separated as superior strains. Original strains ZM2 and ZY2 were breeded by protoptast induced mutation and DES mutagenesis. Pure cultured starters processing technology was achieved using optimum design of breeding technology by RSM. This write up provided a set of processing technique with the maintenance of general quality of traditional fermented barley wine which has subtle taste, low alcohol and typical flavour.

The main results are as following:

1. Quality and flavor characteristics of the barley wine and its starter were identified by analyzing Physiochemical Indexes, GC-MS and sensory index, comparing with Chinese Rice Wine and Japanese sake.(1) Tibet barley wine starter was identified as Chinese yeast with different forms and qualities. The traditional

barley wine was identified as fermentation wine and its physiochemical indexes of were between Rice Wine and Sake. The total content of amino acid in Quality Highland Barley Wine was 110.912 mg/100ml and the wine is also rich in vitamin B_10B_20C and minerals.

2. There are 48 flavor components in barley wine including 16 kinds of alcohol,5 kinds of ester, 6 kinds of acid,10 kinds of aldehyde,7 kinds of ketone and orther phenols. GC-MS shows that: highest content/threshold is aldehyde 41.76,then acetic acid 40.38,Isoamyl alcohol 13.38,caproic acid 4.65, ethyl lactate 3.07,Butyric acid 2.94,ethyl acetate 2.65.â-phenylethanol 2.14,3-hydroxy butanone1.36, these 9 compounds are most responsible for the flavor.2. Microorganisms of Barley wine starter and microorganisms flora during fermentation were studied using routine isolation, purification and identification method together with molecular biological technique, including PCR-Clone Analysis.

It shows that:

1. The identification of dominant mold shows:there were 130 dominant mold strains, including 81 strains of Rhizopus.SPP:Rhizopus oryzae 39 strains, Rhizopus chinensis 15 strains,Rhizopus arrhizus 9 strains, 39 strains of Mucor SPP: including Mucor javanicus 18 strains,Mucor racermosus 11 strains, Mucor rouxians 8 strains, Absidia spp 8 strains.ZM2 was screened from all these mold strains with glucoamylase activity 520.27mg/h·g. ZM2 is indentified as Rhizopus oryzae through ITS sequence analysis and Phylogenetic Tree Construction.
2. The identification of dominant yeast shows:there are dominant yeast 580 strains,belonging to 6 species of Saccharomyces, Saccharomycopsis, Candida, Pichia, Debaryomyces, Cryptococcus, including 186 strains of Saccharomyces cerevisiae,Candida sake 38 strains, Pichia farinosa 16 strains, Debaryomyces hansenii 12 strains, Cryptococcus terreus 5 strains. ZY2 was screened from all these yeast strains for its high alcohol production ability,while ZY28 strain was screened by its ability to produce glucoamylase and good flavour. ZY2 and ZY28 are identified as Saccharomyces boulardii and Saccharomycopsis fibuligera, respectively through ITS sequence analysis and Phylogenetic Tree Construction.
3. The identification of dominant lactic acid bacteria shows:38 strains in 3 species were dominant lactic acid bacteria, including Streptococcus faecalis, Pediococcus pentosaceus, Pediococcus acidialactici, Lactobacillus. johnsonii and Lactobacillus sake. ZB12 strain was screened for further research. It was identified as Pediococcus pentosaceus through ITS sequence analysis and Phylogenetic Tree Construction. 3.ZY2 and ZM2 were breeded by protoplast induced mutation with UV and DES combination.

It shows that:

1. UV-Y2 was obtained with higher alcohol production by UV irradiation of ZY2 protoplast for 50s. The protoplast of UV-Y2 was treated with 1.5% DES for 30

minutes, then the final mutant strain UV-DES-Y2 was obtained which can produce more than 33.33% alcohol after a series of screening by TTC plating screen, Durham fermentation and erlenmeyer flask fermentation.etc.

2. UV-M4 was obtained with higher alcohol production by UV irradiation of ZM2 protoplast for 30s. The protoplast of UV-Y2 was treated with 0.9% DES 50 minutes, then the final mutant strain UV-DES-M4 was obtained which can produce more than 50.9%Glucoamylase.The optimum of enzymatic production for starter 1 is achieved by Using RSM: highland barley powder with 40% water, 8.06% soya flour, 1.0% inoculation, cultivating at 30.05! for 42h. and the Glucoamylase is 895.32 mg/h·g dry starter.
3. The optimal processes of making yeast solid starter: highland barley powder 50% glucose-bean sprout liquid medium (GBM) whose initial pH is 3.5 (for ZY2) or 5 (for ZY28), 15% inoculating ratio which were previously breeded in GBM for 24h, cultivating at 32! (for ZY2)or 25! (for ZY28) for 30h, so the starter 2(ZY2) and starter 3(ZY28) were obtained.
4. The optimization experiments of brewing technique with pure strain fermentation were carried with the following results. Using Tibet barley 320 as raw materials, soaked or no soaked with the ratio of 1:2 water, steamed 2h, with the inoculum's concentration of 2% pure starter 1,2% pure starter 2 and 1% pure starter 3, fermented on the condition of 28 and 72h.
5. he change of compositions of the barley wine with pure strain and the traditional barley wine during its fermentation process are studied. The results were showed as follows: The changing tendency of compositions during two kinds of fermentation process was similar: pH declined, temperature increase slowly and declined slowly with the highest temperature 36!.The activities of the saccharifying enzymes and the reducing sugar content increased significantly (P"ÿ0.05) during the first period of fermentation to 24h, then decreased gradually. The total sugar content decreased significantly (P"ÿ0.05) during all the fermentation while the alcoholicity increased significantly (P"ÿ0.05).The alcoholicity during the fermentation with pure strain process reached to 10.3% higher than traditional fermentation. Both the ammonia-N and the acid prote in ase increased significantly (P"ÿ0.05) up to 24h of fermentation; and then decreased slowly. The total content of barley wine during the fermentation with pure strain reached to 80.923 mg/100ml without Methionine.

FERMENTATION TECHNOLOGY: COOKING UP NEW THERAPEUTICS

Therapeutic drugs are substances used for health restoration in the treatment of disease, drawing on the utilization of microbial bio-factories. These bio-factories offer the pharmaceutical industry a reliable, powerful and economic source for therapeutics. In order to demonstrate the diversity of applications for bio-factories,

Silvian Shama, Business Development Manager at the Jerusalem plant, looks at examples of therapeutic drugs that are primarily isolated through microbial fermentation.

Chemical Therapeutics

The term chemical therapeutics refers to toxic small organic molecules with a molecular weight of 1000-2000 which are produced naturally by microorganisms in response to stress and are called secondary metabolites. These molecules may be complex in nature, such as heterocyclics, with several chiral centers, large stereospecific rings or a unique conjugated double bond system. Small, cytotoxic molecules are employed in anti-cancer therapy (chemotherapy) and also in the treatment of infectious diseases (antibiotics).

Cancer therapy - The anti-cancer therapeutic activity of small molecule drugs in clinical use is limited by their general toxicity to proliferating cells, including some normal cells. In the lack of specificity and effective targeting, chemotherapy using High Potency Active Pharmaceutical Ingredients (HPAPIs) involves high doses and often severe side effects. Over the last few years the pharmaceutical industry has taken a new approach to targeted drug delivery by conjugating HPAPIs to monoclonal antibodies creating highly specific 'smart bombs' which can be directed to cancer cells through highly expressed antigens. This means that not only are smaller quantities of a cytotoxic drug required, but non-target cells are spared many of the cytotoxic effects. Various HPAPI Antibody Drug Conjugates (ADCs) are currently in clinical trials with major pharmaceutical companies and some are gaining FDA approval for difficult-to-treat indications, mainly in the treatment of cancer.

Example 1 – Wyeth's FDA approved Mylotarg

ADC consists of a recombinant humanized IgG4 kappa antibody, conjugated with a cytotoxic anti-tumor antibiotic, calicheamicin, which is isolated through the fermentation of a bacterium, Micromonospora echinospora subspecies calichensis. The antibody portion of Mylotarg binds specifically to the CD33 antigen, a sialic acid-dependent adhesion protein which is found on the surface of leukemic blasts and immature normal cells of myelomonocytic lineage, but not on normal hematopoietic stem cells.

Treatment of infectious diseases - Antibiotics are substances or compounds that kill or inhibit the growth of microorganisms, mainly bacteria (bactericidal or bacterio static agents, respectively). Many of the antibiotics are sourced from living microorganisms, used either in their natural state or in the case of semi-synthesis, chemically modified. Semi–synthesis is usually used when the precursor molecule is too structurally complex, too costly or too inefficient to be produced by total synthesis. It is also possible that the semi synthetic derivative exhibits better performance than the original bio molecule itself in terms of potency, stability or safety.

Example 1 - Aminoglycosides are produced and isolated from bacteria of the Streptomyces genus, used in their existing form to provide drugs such as Streptomycin, a bactericidal agent and Hygromycin B, a killer of bacteria and fungi.

Biological Therapeutics

The term biological therapeutics refers to large organic molecules, mainly proteins. Proteins are polymers of amino acids arranged in a linear chain and folded into a globular form. The sequence of an amino acids protein is defined by the sequence of a gene, which is encoded in the genetic code. Biological therapeutics are isolated from the native producing microorganisms or, more often, expressed in a heterologous microorganism using recombinant DNA technology (recombinant proteins). Fermentation derived biological therapeutics have been employed in a wide range of medical indications.

Native protein example – Collagenase - Collagenases are a group of enzymes produced by the bacterial pathogen Clostridium genus. Upon infection, Collagenases break the collagen peptide bond, a key component of the animal extracellular matrix. Destruction of the extracellular structures causes severe tissue damage and can facilitate the spread of serious medical conditions such as gaseous gangrene.

Essentially turning an enemy into a friend, native isolated collagenases are employed in a variety of medical indications involving collagen disorders.

Collagenases approved for medical uses are: Santyl® ointment and Iruxol for the removal of dead skin from wounds and burned areas in poorly healing wounds and necroses, and Xiafles Dupuytren's contracture, a condition resulting in the contracture of the fingers into the palm. Other collagenases are in currently in clinical trials for indications such as:

- Frozen shoulder - an inflammation and thickening of the shoulder capsule;
- Cellulite - a condition characterized by dimpling of the skin typically affecting the thighs and buttocks; and
- Lipomas - benign fatty tumors that occur as bulges under the skin.

Some approved products are already on the market such as Genentech's Lucentis (ranibizumab injection), which is a recombinant humanized IgG1 kappa iso type monoclonal antibody fragment designed for intraocular use.

Applications of microbial fermentation technology in new therapeutics are expanding and SAFC offers unique cGMP fermentation-based manufacturing capabilities through its Jerusalem facility, including high potent APIs and biologics. SAFC scientific teams provide a complete range of services, including process evaluation and development, optimization and scale up, technology transfer, manufacturing, analytical testing and regulatory filing to help customers bring new drugs to market faster.

The Development of Deep-tank Fermentation

During World War II the governments of the United States and the United Kingdom approached the largest U.S. chemical and pharmaceutical companies to enlist them in the race to mass produce penicillin, the "wonder drug." One of these companies, Pfizer, already had experience with fermentation techniques, first implemented twenty years earlier to manufacture citric acid. Building on that experience, Pfizer succeeded in producing large quantities of penicillin using deep-tank fermentation. Its success helped make penicillin available to Allied soldiers by the end of the war.

The American Chemical Society designated the development of deep-tank fermentation by Pfizer Inc as a National Historic Chemical Landmark on June 12, 2008 in Brooklyn, New York.

THE PENICILLIN PROBLEM

The story of the accidental discovery of penicillin by Alexander Fleming has been told often. Fleming was already well known for his earlier work on staphylococci when, upon returning from a long holiday in September 1928, he noted that one petri dish contained colonies of Staphylocccus except for clear area contaminated with a fungus that appeared to inhibit bacterial growth. Fleming identified the contaminant as a strain of Penicillium and he found that it killed a host of Gram-positive bacteria, including those that caused scarlet fever, pneumonia, gonorrhea, meningitis, and diphtheria.

Cultivating penicillin proved difficult and the small quantities that could be produced were unstable. Fleming also believed that penicillin would not remain in the human body long enough to kill bacteria. Taken together, these facts convinced Fleming that penicillin would never be an important antibiotic for treating infections. For the next decade, penicillin remained a laboratory curiosity, until Howard Florey and a team of researchers at Oxford University demonstrated its potential life-saving properties. The researchers also discovered a method to purify penicillin and to keep it in a stable form. By 1940, clinical trials were underway and their success led to the hunt for a method to manufacture this "wonder drug." But by then England was at war, making mass production difficult.

Pfizer and Fermentation

The answer begins with a small chemical company that would grow into a pharmaceutical giant.

Charles Pfizer and Charles Erhart, cousins and recent émigrés from Germany, established Charles Pfizer & Company in the Williamsburg neighborhood of Brooklyn in 1849. The company's first product cured a common 19th-century malady, intestinal worms, which were usually treated with santonin, an anti-parasitic so bitter that most people thought the cure worse than the condition. But the cousins, pooling the skills of Pfizer, a chemist, and Erhart, a confectioner, blended santonin with an almond-toffee flavoring to create a candy-flavored medicine that patients would take.

Pfizer's initial success with santonin encouraged the cousins to look for other opportunities to manufacture "fine chemicals." Among the company's products were quinine, used as an analgesic; borax, used as a preservative and a laundry detergent; boric acid, a topical antiseptic; cream of tartar, used in baking powder, and tartaric acid, also used in foods (both by-products of the European wine industry); Rochelle salts, a laxative and diuretic; camphor, a painkiller; and strychnine, a tonic and appetite stimulant. During the Civil War Pfizer, like other chemical and pharmaceutical companies, responded to the demand for painkillers, preservatives, and disinfectants, producing many products with medicinal applications, including iodine, morphine, and chloroform.

Pfizer's initial blockbuster product was citric acid, which it added to its product line in the 1880s. Citric acid is used in foods and beverages – notably soft drinks – because it is a natural preservative which can also add an acidic, slightly sour, taste. By supplying citric acid for new, popular drinks like Coca-Cola, Pepsi-Cola, and Dr. Pepper, Pfizer was able to dominant the market. At first, the company used the standard production methods, obtaining citric acid from unripe fruit from Italy, California, Florida, and the West Indies, and treating it with lime (CaO) and sulfuric acid to yield a solution of crystallized citric acid.

The botanist Carl Wehmer discovered in 1893 that the Penicillium mold could produce citric acid from sugar. Later, J.A. Martin discovered that fermenting sugar could yield citric acid. But these were ideas ahead of their time because no one knew how to manufacture citric acid from these sources on a commercial scale. That is, until James Currie, a food chemist, discovered that citric acid could be fermented from certain strains of the mold Aspergillis niger combined with sugar.

Currie later described how he took his discovery to Pfizer in 1917: "I met with John Anderson, who was Pfizer's chairman of the board. During our first meeting, Mr. Anderson introduced me to another Pfizer official with the remark, 'Dr. Currie is up here now and I think he has something interesting.'" Pfizer officials quickly realized that the "something interesting" was the prospect of producing vast quantities of citric acid from sugar, rather than from imported citrus fruit.

The fact that Currie joined Pfizer during World War I was no coincidence. Pfizer officials, led by John Anderson, realized that the company had to reduce its vulnerability as an importer of raw materials. The war cut off the supply of Italian citrus, and all raw material deliveries from Europe were threatened by the German submarine blockade. Supplying medicines and chemicals for the war helped the company survive, but overall sales declined. Now, more than ever, Pfizer needed to find ways to produce citric acid without using citrus.

Currie, aided by his precocious 16-year-old lab assistant Jasper Kane, tackled the fermentation problem. He knew Aspergillis niger is aerobic, meaning it needs air to grow. Currie tried to grow the mold in a large flat pan purchased at the five-and-ten, but had limited success. He cut the pan into smaller, shallower pans, and immediately increased the yield. Still, the process was subject to a number of variables: the quality of the mold spores, the purity of the cultures, contamination of air and the medium, humidity and temperature, and many others.

Currie plugged on, turning to Gordon Dryer machines, which controlled humidity and temperature. He also experiment with different-sized pans and their depth and with temperatures. In 1919 Pfizer opened a pilot plant using his fermentation process, named SUCIAC – "Sugar Under Conversion to Citric Acid." Still, yields were disappointingly low for a time, so Currie kept fine tuning the process – using smaller pans and a better ventilation system – to increase output. By 1924 growing yields of citrus acid convinced Pfizer officials to build a SUCIAC plant. The new structure began operating in 1926; that year the output of citric acid using fermentation technology far outpaced the production based on lemons and limes. By the end of the decade Pfizer was dominating the market and in 1929 the company reached a milestone; it produced 5.9 million pounds of citric acid without using any lemon juice, lime juice, or foreign citrate of lime in its production process.

Deep-tank Fermentation

Pfizer's experience with fermentation in the production of citric acid provided valuable lessons for the company when biochemistry created new germ-killing antibiotics, many of which could be produced in the laboratory but not on a commercial scale. And Pfizer gained even more experience as the company searched for ways to use fermentation to produce other chemicals or to improve existing methods. For example, in 1933, Jasper Kane substituted molasses, a cheap by-product of sugar refining, for sugar in the fermentation of citric acid.

Pfizer soon was producing other products by fermentation, most notably gluconic acid (C6H12O7), used as an additive in food to regulate acidity and as a cleaning agent. Unlike the case of citric acid, an important product that Pfizer sought to produce more efficiently, Pfizer succeeded in marketing gluconic acid only after developing a commercial process involving deep-tank fermentation. Pfizer had tried unsuccessfully to use deep tanks to produce citric acid; in 1929 the company succeeded in producing gluconic acid in a submerged aerobic medium in stirred deep tanks which controlled for pH and the sterility of the air.

In the following years the company used fermentation processes to produce itaconic acid, fumaric acid, and intermediates for chemical synthesis. The production of gluconic acid and other products using deep-tank fermentation was a significant breakthrough for Pfizer. While the company used fermentation techniques in manufacturing citric acid, it was the development of deep tanks for gluconic acid that paved the way for the successful mass production of penicillin.

In the 1930s Pfizer, which had close ties with the pharmaceutical and food industries, became interested in the commercial production of vitamins, beginning with vitamin C. Ascorbic acid was first isolated by Albert Szent-Györgyi in 1928; shortly after, its structural formula was established and a method of synthesis discovered. 1936 Pfizer began marketing synthetic vitamin C made with submerged fermentation as the first step in the process. Pfizer soon became the leading manufacturer, and the company decided to market other vitamins, introducing vitamin B2 in 1938 and, after the war, vitamin B12.

Penicillin Solved

In 1941 the British government sought help from the U.S. scientific community in solving the problem of mass producing penicillin. Four chemical and pharmaceutical companies, Pfizer included, responded, plunging into the race to produce the world's first "wonder drug."

In the early 1940s Jasper Kane and his colleagues increased penicillin production incrementally as well as the drug's potency and purity, but the gains were frustratingly slow. At first, Kane used flasks and pans similar to the fermentation techniques for citric acid. In 1942, Kane suggested switching to the deep tanks that worked so well for gluconic acid. Kane's suggestion was risky for it meant Pfizer would have to curtail production of other, more profitable products while it concentrated on penicillin. As one executive, John Smith, said: "The mold is as temperamental as an opera singer, the yields are low, the isolation murder, the purification invites disaster. Think of the risks!"

The company decided to take the gamble. In September 1943 it purchased the old Rubel Ice Plant, a nearby building that had the refrigeration equipment required, rebuilding it into the world's first large-scale penicillin factory. On March 1, 1944 Pfizer's penicillin plant opened. It contained fourteen 7,500-gallon tanks; soon the company was producing five times more penicillin then originally estimated, making Pfizer the leading supplier of the drug; most of the penicillin that went ashore with Allied forces on D-Day came from this plant on Marcy Avenue in Brooklyn.

Behind the numbers stood a marvel of chemical engineering. Production began with a sterile culture of the penicillin mold, which then was propagated, first in three-liter flasks, next in 200-gallon "seed" tanks. The culture then moved to huge fermenter tanks containing microbe fodder, chiefly corn steep liquor, milk sugar, salts, and minerals. The mold was allowed to grow for two to four days.

The trickiest part (the whole process was tricky) was extracting penicillin from the broth, for penicillin was a very stingy "magic bullet:" only four parts drug per 10,000 parts broth. The extracted material was then purified and bottled in sterile rooms as extreme care had to be taken in these last steps to avoid contamination. After the bottles were filled, the penicillin moved through a freezing apparatus and then into a vacuum drier which dehydrated the drug.

This was the process that helped win World War II. Later, Pfizer improved the procedure. In 1946, for example, Pfizer researchers discovered that the normal yellow color of penicillin indicated the presence of impurities. The scientists developed a crystallization method that yielded white penicillin stable at room temperature and potent for years.

Pfizer used the lessons learned to develop other antibiotics, first applying its fermentation techniques to the manufacture of streptomycin, a drug developed by

Selman Waksman and his staff at Rutgers. In 1949 Pfizer scientists, after testing tens of thousands of soil samples, found a micro-organism in soil from America's Midwest that proved effective against a wide range of deadly bacteria. Terramycin , derived from the Latin for "earth fungus," was the first antibiotic developed exclusively by Pfizer scientists.

Alexander Fleming and the discoverers of other antibiotics deserve all the credit they have won. But their work would have remained on the laboratory shelf if not for the development of deep-tank fermentation. As David Wilson wrote in his book In Search of Penicillin (Knopf, 1976), "It is the biggest single failing of the myth about penicillin that it ignores the technological breakthrough of deep fermentation, a breakthrough that was every bit as vital to the successful development of penicillin as any of the more dramatic laboratory work."

In the early 20th century Pfizer developed innovative fermentation technology, applying it first to the mass production of citric acid. In subsequent years, under the direction of James Currie and Jasper Kane, Pfizer perfected deep-tank fermentation, an aseptic process for growing large quantities of microorganisms which require oxygen for survival. When scientists in England were unable to produce penicillin on a large scale during World War II, Kane suggested trying deep-tank fermentation. In a major feat of chemical engineering, the company rebuilt an old ice plant, which had the refrigeration machinery required for submerged fermentation, and opened the world's first large-scale penicillin facility on March 1, 1944. Pfizer manufactured other antibiotics, notably Terramycin, and vitamins using deep-tank fermentation techniques.

5

INDUSTRIAL MICROBIOLOGY

'Industrial microbiology or microbial biotechnology encompasses the use of micro-organisms in the make the product by machine of meal or industrial products. The use of micro-organisms for the production of meal, either human or animal, is often considered a branch of meal microbiology. The microorganisms used in industrial processes may be natural isolates, laboratory selected mutants or genetically engineered organisms.

FOOD MICROBIOLOGY

Youghourt, cheese, chocolate, butter, pickles, sauerkraut, soya sauce, vitamins, amino acids, meal thickeners (microbial polysaccharides), alcohol, sausages, and silage (animal meal) are all produced by industrial microbiology processes. 'Good' bacteria such as probiotics are becoming increasingly important in the meal industry.

BIOPOLYMERS

The huge is a type of biopolymers as polysaccharides, polyesters, and polyamides, are produced by micro-organisms. These products range from viscous solutions to plastics. The genetic manipulation of microorganisms has permitted the biotechnological production of biopolymers with tailored material properties suitable for high-value medical application such as tissue engineering and drug delivery. Industrial microbiology can be used for the biosynthesis of xanthan, alginate, cellulose, cyanophycin, poly (gamma-glutamic acid), levan, hyaluronic acid, organic acids, oligosaccharides and polysaccharides, and polyhydroxyalkanoates by jose landron

BIOREMEDIATION

This is used to different environment like that clean-up contaminated a surrounding object. These bioremediation and biotransformation methods harness naturally occurring microbes to degrade, transform or accumulate a huge range of compounds adding hydrocarbons polychlorinated biphenyls (PCBs), polyaromatic hydrocarbons (PAHs), pharmaceutical substances, radionuclides and metals.

WASTE BIOTREATMENT

Microorganisms are used to treat the vast quantities of wastes generated by modern societies. Biotreatment, the processing of wastes using living organisms, is a surrounding objects friendly, relatively simple and cost-effective alternative to physico-chemical clean-up options. Confined a surrounding objects, such as bioreactors can be employed in biotreatment processes.

Wastewater Treatment

Biological wastewater manner of applying remedies is undoubtedly one of the most important biotechnological processes, which have been used for over a century to treat municipal and industrial wastewaters. Culture-independent molecular techniques have been used to study the diversity and physiology of ecologically important micro_organisms in wastewater manner of applying remedies processes. A number of new exciting insights into the structure, function, and dynamics of complex microbial communities in wastewater manner of applying remedies processes have been gained, which have significantly expanded our understanding of process design, operation and control.

HEALTH CARE AND MEDICINE

Micro-organisms can help in the manner of applying remedies of illness such as cancer. Research shows that clostridia can selectively target cancer cells. Various strains of non-pathogenic clostridia have been shown to infiltrate and replicate within solid tumours. Clostridia so have the potential to deliver therapeutic proteins to tumours. Lactobacillus spp. and other lactic acid bacteria possess numerous potential therapeutic properties adding anti-inflammatory and anti-cancer activities.

CORYNEBACTERIA

Corynebacteria are a diverse group Gram-positive bacteria found in a range of different ecological niches such as soil, vegetables, sewage, skin, and cheese smear. Corynebacterium glutamicum is of immense industrial momentous and is one of the biotechnologically most of great consequence bacterial species with an annual production of more than two million tons of amino acids, mainly L-glutamate and L-lysine. The genome sequence of C. glutamicum has been published.

XANTHOMONAS

The genus Xanthomonas consists of 20 plant-associated species, many of which cause momentous illness of crops and other plants. The particular species comprise multiple pathovars, characterised by distinctive host specificity or mode of infection. Bacteria of the genus Xanthomonas are able to produce the acidic exopolysaccharide xanthan gum. Because of its physical properties, it is widely used as a viscosifer, thickener, emulsifier or stabilizer in both meal and non-meal industries.

The types of Industrial Processes

Industrial processes are procedures including chemical or mechanical steps to aid in the make the product by machine and other of an item or items, usually carried out on a very large scale. Industrial processes are the key components of heavy industry.

Most processes make the production of an otherwise rare material vastly cheaper in price, thus changing it into a commodity; the process makes it economically feasible for society to use the material on a large scales, in machinery, or a substantial amount of raw materials, in comparison to batch or craft processes. Production of a specific material may involve more than one type of process. Most industrial processes result in both a desired product(s) and by-products, many of which are toxic, hazardous, or hard to deal with. Very, very few processes are self-contained.

The part of a larger process are given.

1.Liquefaction of gases—for ease of transportation.

2.Supercritical drying, Freeze drying—removal of excess liquid.

3.Scrubber - removing of pollution from exhaust gases.

LIQUEFACTION OF GASES

Liquefaction of gases adds a number of phases used to convert a gas into a liquid state. The processes are used for scientific, industrial and commercial purposes. Many gases can be put into a liquid state at normal atmospheric pressure by simple cooling; a few, such as carbon dioxide, require pressurisation as well. Liquefaction is used for analyzing the fundamental properties of gas molecules, for storage of gases, for e.g.: LPG, and in refrigeration and air conditioning. There the gas is liquefied in the condenser, where the heat of vaporisation is released, and evaporated in the evaporator, where the heat of vaporisation is absorbed. Ammonia was the first such refrigerant, but it has been replaced by compounds derived from petroleum and halogens.

Liquefaction of helium (4He) with the Hampson-Linde cycle led to a Nobel Prize for Heike Kamerlingh Onnes in 1913. At ambient pressure the boiling point of liquefied helium is 4.22 K (-268.93°C).2.17 K liquid 4He has many amazing properties, such as climbing the walls of the vessel, exhibiting zero viscosity, and offering no lift to a wing past which it flows.

The liquefaction of gases is a complicated process that uses various compressions and expansions to achieve high pressures and very low temperatures; using for example turboexpanders.

The liquefaction of air is used to obtain nitrogen, oxygen and argon by separating the air components by distillation.

SUPERCRITICAL DRYING

Supercritical drying is a process to remove liquid in a precisely controlled way, similar to freeze drying. It is useful in the production of micro electro mechanical systems (MEMS), in the drying of spices, is commonly used in the production of aerogel and in the preparation of biological specimens for scanning electron microscopy.

To avoid this, the sample can be brought from the liquid phase to the gas phase without crossing the liquid-gas boundary on the phase diagram; in freeze-drying, this means going around to the left (low temperature, low pressure). Above all some structures are disrupted even by the solid-gas boundary. Supercritical drying, on the other hand, goes around the line to the right, on the high-temperature, high-pressure side. This route from liquid to gas does not cross any phase boundary, instead passing through the supercritical region, where the distinction between gas and liquid ceases to apply.

Fluids suitable for supercritical drying include carbon dioxide and freon. Nitrous oxide has similar physical behaviour to carbon dioxide, but is a powerful oxidizer in its supercritical state. Supercritical water is also a powerful oxidizer, partly because its critical point occurs at such a high temperature (374 °C) and pressure.

Food and Food additives

Meal additives are substances added to meal to preserve flavour or improve its taste and appearance. Some additives have been used for centuries; for e.g., preserving meal by pickling (with vinegar), salting, as with bacon, preserving sweets or using sulfur dioxide as in some wines. With the advent of processed meals in the second half of the 20th century, many more additives have been introduced, of both natural and artificial origin.

Numbering

To regulate these additives, and inform consumers, each additive is assigned a unique number. Initially these were the "E numbers" used in Europe for all approved additives. This numbering scheme has now been adopted and extended by the Codex Alimentarius Commission to internationally identify all additives, regardless of whether they are approved for use.

E numbers are all prefixed by "E", but countries outside Europe use only the number, whether the additive is approved in Europe or not. For e.g., acetic acid is written as E260 on products sold in Europe, but is simply known as additive 260 in some countries. Additive 103, alkanet, is not approved for use in Europe so does not have an E number, although it is approved for use in Australia and New Zealand. Since 1987 Australia has had an approved system of labelling for additives in packaged meals. Each meal additive has to be named or numbered. The numbers are the same as in Europe, but without the prefix 'E'.

Acids

Meal acids are included to make flavours "sharper", and also act as preservatives and antioxidants. Common meal acids include vinegar, citric acid, tartaric acid, malic acid, fumaric acid, lactic acid.

Acidity regulators

Acidity regulators are used to change or otherwise control the acidity and alkalinity of meals.

Anticaking agents

Anticaking agents keep powders such as milk powder from caking or sticking.

Antifoaming agents

Antifoaming agents reduce or prevent foaming in meals.

Antioxidants

Antioxidants such as vitamin C act as preservatives by inhibiting the effects of oxygen on meal, and can be beneficial to health.

Bulking agents

Bulking agents such as starch are additives that increase the bulk of a meal without affecting its nutritional value.

Meal colouring

Colourings are added to meal to replace colours lost during preparation, or to make meal look more attractive.

Colour retention agents

In contrast to colourings, colour retention agents are used to preserve a meal's existing colour. Colour retention agents are often used for restoring cow's milk to its natural white colour as unhealthy cows may bleed into the milk.

Emulsifiers

Emulsifiers allow water and oils to remain mixed together in an emulsion, as in mayonnaise, ice cream, and homogenised milk.

Flavours

Flavours are additives that give meal a particular taste or smell, and may be derived from natural ingredients or created artificially.

Flavour enhancers

Flavour enhancers enhance a meal's existing flavours. They may be extracted from natural sources (through distillation, solvent extraction, maceration, among other methods) or created artificially.

Flour treatment agents

Flour manner of applying remedies agents are included to flour to improve its colour or its use in baking.

Humectants

Humectants prevent meals from drying out.

Tracer gas

Tracer gas allow for package integrity testing to prevent meals from being exposed to atmosphere, thus guaranteeing shelf life.

Preservatives

Preservatives prevent or inhibit spoilage of meal due to fungi, bacteria and other micro-organisms.

Stabilisers

Stabilisers, thickeners and gelling agents, like agar or pectin (used in jam for example) give meals a firmer texture. While they are not true emulsifiers, they help to stabilise emulsions.

Sweeteners

Sweeteners are included to meals for flavouring. Sweeteners other than sugar are included to keep the meal energy (calories) low, or because they have beneficial effects for diabetes mellitus and tooth decay and diarrhea.

Thickeners

Thickeners are substances which, when added to the mixture, increase its viscosity without substantially modifying its other properties.

SAFETY

With the increasing use of processed meals since the 19th century, there has been a great increase in the use of meal additives of varying levels of safety. This has led to legislation in many countries regulating their use. For e.g., boric acid was widely used as a meal preservative from the 1870s to the 1920s, but was banned after World War I due to its toxicity, as demonstrated in animal and human studies. During World War II the urgent need for cheap, available meal preservatives led to it being used again, but it was finally banned in the 1950s. Such cases led to a general mistrust of meal additives, and an application of the precautionary principle led to the conclusion that only additives that are known to be safe should be used in meals. In the USA, this led to the adoption of the Delaney clause, an amendment to the Federal Meal, Drug, and Cosmetic Act of 1938, stating that no carcinogenic substances may be used as meal additives. Above all, after the banning of cyclamates in the USA and Britain in 1969, saccharin, the only remaining legal artificial sweetener at the time, was found to cause cancer in rats. Widespread public outcry in the USA, partly communicated to Congress by postage-paid postcards supplied in the packaging of sweetened soft drinks, led to the retention of saccharin despite its violation of the Delaney clause.

In September 2007, research financed by Britain's Meal Standards Agency and published online by the British medical journal The Lancet, presented evidence that a mix of additives commonly found in children's meals increases the mean level of hyperactivity. The team of researchers concluded that "the finding lends strong support for the case that meal additives exacerbate hyperactive behaviours (inattention, impulsivity and overactivity) at least into middle childhood." That study examined the effect of artificial colours and a sodium benzoate preservative, and found both to be problematic for some children. Further studies are needed to find out whether there are other additives that could have a similar effect, and it is unclear whether some disturbances can also occur in mood and concentration in some adults.

"Although quite complicated, this was a carefully conducted study in which the investigators went to great lengths to eliminate bias and to rigourously measure outcomes. The results are hard to follow and somewhat inconsistent. For many of the assessments there were small but statistically significant differences of measured behaviours in children who consumed the meal additives compared with those who did not. In each case increased hyperactive behaviours were associated with consuming the additives. For those comparisons in which no statistically significant differences were found, there was a trend for more hyperactive behaviours associated with the meal additive drink in virtually every assessment. Thus, the overall findings of the study are clear and require that even we skeptics, who have long doubted parental claims of the effects of various meals on the behaviour of their children, admit we might have been wrong."

Baker's yeast

Baker's forth appearing in fermentation is the common name for the strains of yeast commonly used as a leavening agent in baking bread and related products, where it converts the fermentable sugars present in the dough accelerates the growth of yeasts. Salt and fats such as butter slow down yeast growth. The majority of the forth appearing in fermentation used in baking is Saccharomyces cerevisiae, which is the same species commonly used in alcoholic fermentation, and so is also called brewer's yeast. Additionally, Saccharomyces exiguus is a wild forth appearing in fermentation found on plants, fruits, and grains that is occasionally used for baking; it is not, above all, generally used in a pure form, but comes from being propagated in a sourdough starter.

The first records that show this use came from Ancient Egypt. Researchers speculate that a mixture of flour meal and water was left longer than usual on a warm day and the yeasts that occur in natural contaminants of the flour caused it to ferment before baking. The resulting bread would have been lighter and more tasty than the previous flat, hard flatbreads. It is generally assumed that the earliest forms of leavening were likely very similar to modern sourdough; the leavening action of forth appearing in fermentation would have been discovered from its action on flatbread doughs, and would either have been cultivated separately or transferred from batch to batch by means of previously mixed dough. Alternately, the development of leavened bread seems to have developed in close proximity to the development of beer brewing, and barm from the beer fermentation process can also be used in bread making.

Without a full understanding of microbiology, early bakers would have had little ability to directly control yeast cultures, but still kept locally interesting cultures by reusing doughs and starters to leaven later batches. Above all, it became possible to isolate and propagate favoured yeast strains in the same manner as was done in the beer industry, and it eventually became practical to propagate forth appearing in fermentation in a slurry with a composition similar to beer wort, usually including barley malt and wheat flour. Such cultures (sometimes referred to in old American cookery as "emptins", from their origins as the dregs of beer or cider fermentation) would become the ancestors of modern baker's forth appearing in fermentation, as they generally were carefully maintained to avoid what would later be discovered to be bacterial contamination, including using preservatives such as hops as well as boiling the growth medium.

Later refinements in microbiology following the work of Louis Pasteur led to more advanced methods of culturing pure strains, making modern commercial forth appearing in fermentation possible, and turning yeast production into a major industrial endeavor. The slurry forth appearing in fermentation made by small bakers and grocery shops became cream yeast, a suspension of live yeast cells in growth medium, and then compressed yeast, the fresh cake yeast that became the standard leaven for bread bakers in much of the Westernised world during the early-20th century.

TYPES OF BAKER'S YEAST

Baker's yeast is available in a number of different forms. Though each version has certain advantages over the others, the choice of which form to use is largely a question of the requirements of the recipe at hand and the training of the cook preparing it. With occasional allowances for liquid content and temperature, the different forms of commercial forth appearing in fermentation are given.

1.Cream yeast is the closest form to the yeast slurries of the 19th century, being essentially a suspension of yeast cells in liquid, siphoned off from the growth medium. Its primary use is in industrial bakeries with special high-volume dispensing and mixing equipment, and it is not readily available to small bakeries or home cooks.

2.Compressed yeast is essentially cream yeast with most of the liquid removed. It is best known in the form of cake yeast, which is essentially a soft solid, beige in colour, but is also available in crumbled form for bulk usage. It is highly perishable; though formerly widely available for the consumer market, it has become less common in supermarkets in some countries due to its poor keeping properties, having been superseded in some such markets by active dry and instant yeast. It is still widely available for commercial use, and is somewhat more tolerant of low temperatures than other forms of commercial yeast; above all, even there, instant forth appearing in fermentation has made significant market inroads.

3.Active dry yeast is the form of yeast most commonly available to non-commercial bakers, as well as the forth appearing in fermentation of choice for situations where long travel or uncontrolled storage conditions are likely. It consists of coarse oblong granules of yeast, with live yeast cells encapsulated in a thick jacket of dry, dead cells with some growth medium. Under most conditions, active dry yeast must be proofed or rehydrated first and, despite its better keeping qualities than other forms, is generally considered more sensitive than other forms to thermal shock when actually used in recipes. Active dry yeast also provides an alternative to butter and salt for seasoning pop corn.

4.Instant forth appearing in fermentation appears similar to active dry yeast, but has smaller granules with substantially higher percentages of live cells. It is more perishable than active dry yeast, but also does not require rehydration, and can usually be added directly to all but the driest doughs. Instant yeast generally has a small amount of ascorbic acid included as a preservative. Some producers provide two or more forms of instant forth appearing in fermentation in their product portfolio; for example,

LeSaffre's "SAF Instant Gold" is designed specifically for doughs with high sugar contents.

1.Rapid-rise yeast is a variety of yeast (usually a form of instant yeast) designed to provide greater carbon dioxide output to allow faster rising at the expense of shortened fermentation times. There is considerable debate as to the value of such a product; while most baking experts believe it reduces the flavour potential of the finished product, Cook's Illustrated magazine, among others, feels that at least for direct-rise recipes, it makes little difference. Rapid-rise yeast is often marketed specifically for use in bread machines.

2.Flake yeast is dead yeast, sold primarily as a nutritional supplement. It has little to no leavening power.

For most commercial uses, yeast of any form is packaged in bulk (blocks or freezer bags for fresh yeast; vacuum-packed brick bags for dry or instant); however, yeast for home use is often packaged in pre-measured doses, either small squares for compressed yeast or sealed packets for dry or instant. A single dose is generally about 2.5 tsp or about 7g (1/4 ounce), the reckoned for the average breacrecipe of between 500 gm and 1000gm of diughs though comparatively lesser amounts are used when the yeast is used in a pre-ferment.

USE IN RESEARCH

It is readily available and easy to culture, baker's yeast has long been used in chemical, biological, and genetic research. In 1996, after 6 years of work, S. cerevisiae became the first eucaryote to have its entire genome sequenced. It has over 12 million base pairs and around 6,000 genes. Since then it has remained on the forefront of genetic research. For e.g., most of our information of the cell division cycle was worked out from experiments with forth appearing in fermentation.

Baker's yeast contains enzymes which can reduce a carbonyl group into a hydroxyl group in fairly high submit, thus making it a useful bio-reagent in chemical syntheses. It is known to reduce organometallic carbonyl compounds in very high yield.

Baker's yeast can also be used to produce ethanol via fermentation for use in chemical synthesis, although doing so in some places requires permits.

Reduction of a carbonyl to a hydroxyl with baker's yeast.

Food and Fodder Yeast

Forth appearing in fermentation propagated essentially for meal purposes is known as meal yeast. Yeast produced chiefly to feed animals is called fodder forth appearing in fermentation. The mass cultivation of yeast for use as meal, more specifically as a meal supplement, is to compensate for the dietary inadequacies of cheap meal materials, especially in the regions of the world where human malnutrition is chronic Secondly, there is great emphasis and interest or lowering the BOD of the effluents from industrial plants. Thus several processes to manufacture yeast from industrial and agricultural wastes have been undertaken. This converts waste into products of value, and at the same time prevents environmental pollution.

Single cell protein

Single cell protein (SCP) typically refers to sources of mixed protein extracted from pure or mixed cultures of algae, forth appearing in fermentation, fungi or bacteria used as a substitute for protein-rich meals, in human and animal feeds.

HISTORY

Since 2500 BC yeasts have been used in bread and beverage production. In 1781 processes for preparing highly concentrated forms of forth appearing in fermentation were established. In 1919 Endomyces vernalis submited fats from sulphite liquor (from paper manufacture), and similarly in 1941 employing Geotrichum. The term SCP was coined in 1966 by Carol L. Wilson at MIT

SCP PRODUCTION PROCESS

Single cell proteins develop when microbes ferment waste materials (including wood, straw, cannery and meal processing wastes, residues from alcohol production, hydrocarbons, or human and animal excreta). The problem with extracting single cell proteins from the wastes is the dilution and cost. They are found in very low concentrations, usually less than 5 per cent. Engineers have developed ways to increase the concentrations adding centrifugation, flotation, precipitation, coagulation and filtration, or the use of semi-permeable membranes.

The single cell protein needs to be dehydrated to approximately 10 per cent moisture content and/or acidified to aid in storage and prevent spoilage. The methods to increase the concentrations to adequate levels, and de-watering process require equipment that is expensive and not always suitable for small-scale operations. It is economically prudent to feed the product locally and shortly after it is produced.

Microbes employed include yeasts (Saccharomyces cerevisiae, Candida utilis=Torulopsis and Geotrichum candidum (=Oidium lactis)), other fungi (Aspergillus oryzae, Sclerotium rolfsii, Polyporus and Trichoderma), bacteria (Rhodopseudomonas capsulata), and algae (Chorella and Spirulina). Typical yields of 43 to 56 per cent, with protein contents of 44 to 60 per cent.

The fungus Scytalidium acidophilum grows at below pH 1, offering advantages of low-cost aseptic conditions, avoiding over 100-fold dilution of the acidic hydrolysates to pH values needed for other microbes, and after the biomass is harvested, the acids can be reused.

Commercial production of SCP includes Cyanotech in Hawaii and Earthrise in California.

Meal waste

Meal waste is "any meal substance, raw or cooked, which is discarded, or intended or required to be discarded", according to the legal definition of waste by the EU Commission. Since there are several definitions of waste, equally many definitions of meal waste exist; professional bodies, adding international organisations, state governments and secretariats may formally have their own defintions.

DEFINITION

The definition of waste is a contended subject, often defined on a situational basis, so it follows that meal waste is the same; professional bodies, including international organisations, state governments and secretariats may formally have their own definitions.

In 1975, the European Commission, the executive branch of the European Union, legally defined waste for countries in the Union as: "any substance or object which the holder disposes of or is required to dispose of pursuant to the provisions of national law in force". This directive (75/442/EEC) was amended in 1991 (by 91/

156), with the addition of "categories of waste" and the omission of any reference to national law. Categorises types of waste by how they occur, and some categories appear specific to certain waste types; "Products whose date for appropriate use has expired" targets meal waste, with "date" referring to the expiry date of a meal.

The United States Environmental Protection Agency defines meal waste for the United States as being: "Uneaten meal and meal preparation wastes from residences and commercial establishments such as grocery stores, restaurants, and produce stands, institutional cafeterias and kitchens, and industrial sources like employee lunchrooms". Although it is a nation-wide agency, states are free to define meal waste individually, according to policies, preference and other definitions, though many choose not to.

Meal Processing

Meal waste continues post-harvest, but the amounts involved are relatively unknown and difficult to estimate. Regardless, the variety of factors that contribute to meal waste, both biological/environmental and socio-economical, would limit the usefulness and reliability of general figures. In storage, considerable quantitative losses can be attributed to pests and microorganisms. This is a particular problem for countries that experience a combination of heat (around 30°C) and ambient humidity (between 70 and 90 percent), conditions that encourage the reproduction of insect pests and microorganisms. Losses in the nutritional value, caloric value and edibility of crops, by extremes of temperature, humidity or the action of microorganisms, also account for meal waste; these "qualitative losses" are more difficult to assess than quantitative ones. Further losses are generated in the handling of meal and by shrinkage in weight or volume.

Some of the meal waste produced by processing can be difficult to reduce without affecting the quality of the finished product. Meal safety regulations are able to claim meals which contradict standards before they reaches markets. Although this can conflict with efforts to reuse meal waste (such as in animal feed), safety regulations are in place to ensure the health of the consumer; they are vitally momentous, especially in the processing of mealstuffs of animal origin (e.g., meat and dairy products), as contaminated products from these sources can lead to and are associated with microbiological and chemical hazards.

In the UK, 6.7 million tonnes per year of wasted meal (purchased and edible meal which is discarded) amounts to a cost of £10.2 billion each year. This translates a cost of £250 to £400 a year for every British household.

The per capita annual meal waste output in North America in 1918 was estimated as 100-200 pounds, with the largest source from domestic properties, although that may have little relevance today.

A study by the University of Arizona in 2004, indicated that 14-15 per cent of US edible meal is untouched or unopened, amounting to $43 billion worth of discarded, but edible, meal.

Collection

In areas where waste collection is a public function, meal waste is usually managed by the same governmental organisation as other waste collection. Most meal waste is combined with general waste at the source. Separate collections have the advantage that meal wastes can be disposed of in ways not applicable to other wastes.

From the end of the 19th century through the middle of the twentieth century, many municipalities collected meal waste separately. This was typically disinfected by steaming and fed to pigs, either on private farms or in municipal piggeries.

Separate carbazide collection of meal waste is now being revived in some areas. To keep collection costs down and raise the rate of meal waste segregation, some local authorities, especially in Europe, have introduced "alternate weekly collections" of biodegradable waste which enable a wider range of recyclable materials to be collected at reasonable cost, and improve their collection rates. Above all, they result in a two week wait before the waste will be collected. So there is criticism that, particularly during hot weather, meal waste rots and stinks, and attracts vermin.

Much kitchen waste also leaves the home through garbage disposal units. Dumping meal waste in a landfill causes a surrounding objects damage. By volume, it is the largest contributor to methane gas production. It causes odour as it decomposes, attracts flies and vermin, and has the potential to add biological Oxygen Demand (BOD) to the leachate. The EU Landfill Directive and Waste Regulations, like regulations in other countries, enjoin diverting organic wastes away from landfill disposal for these reasons.

Meal waste can be composed at home, avoiding central collection entirely, and many local authorities have schemes to provide subsidised composting bin systems. Above all, the proportion of the population willing to dispose of their meal waste in that way may be limited.

Anaerobic digestion produces both useful gaseous products and a solid fibrous "compostable" material. Anaerobic digestion plants can provide energy from waste by burning the methane created from meal and other organic wastes to generate electricity, defraying the plant's costs and reducing greenhouse gas emissions.

Alcoholic beverage

An alcoholic beverage is a drink that contains ethanol. Alcoholic beverages are divided into three general classes:

(i) beers.
(ii) wines.
(iii) spirits.

Alcoholic beverages are consumed in almost every sovereign state, and most have laws that regulate their production, sale, and consumption. In particular, such laws specify the minimum age at which a person may legally buy or drink alcoholic beverages. This minimum age can be as low as 16 years in some nations, above all most nations set the minimum age at 18 years.

The production and consumption of alcohol occurs nowadays in most cultures of the world, from hunter-gatherer peoples to nation-states. Alcoholic beverages are often an momentous part of social events in these cultures. In many cultures, drinking plays a significant role in social interaction — mainly because of alcohol's neurological effects.

Alcohol is a psychoactive drug that has a depressant effect. A high blood alcohol content is usually considered to be legal drunkenness because it reduces attention and slows reaction speed. Alcoholic beverages can be addictive, and the state of addiction to alcohol is known as alcoholism.

TYPES

There are many types of Alcoholic beverage

BEER

Beer is mostly used as allover world, this is alcoholic beverage and also famous drink over all after with other things (related to food) It is produced by the brewing and fermentation of starches which are mainly derived from cereal grains—most commonly malted barley although wheat, maize (corn), and rice are also used. Alcoholic beverages which are distilled after fermentation, fermented from non-cereal sources such as grapes or honey, or fermented from unmalted cereal grain, are not classified as beer.

Beer is part of the culture of various nations and has acquired social traditions such as beer festivals and pub culture, which involves activities such as pub crawling and pub games.

The basics of brewing beer are shared across national and cultural boundaries. The two main types of beer are lager and ale, which is further classified into varieties such as pale ale, stout, and brown ale. The beer-brewing industry is a global business, consisting of several dominant multinational companies and thousands of smaller producers, which range from brewpubs to regional breweries.

WINE

Wine involves a longer (complete) chemical is change by fermenting process and a long aging process (months or years) that results in an alcohol content of 9 per cent–16 ABV. Sparkling wine can be made by including a small amount of sugar before bottling, which causes a secondary chemical is change by fermenting to occur in the bottle.

SPIRITS

The spirits is alcoholic beverages and also have 20 per cent ABV Spirits are produced by distillation of a fermented product; this process concentrates the alcohol and eliminates some of the congeners. Spirits can be included to wines to create fortified wines, such as port and sherry. Content of alcohol and liquid without water.

The concentration of alcohol in a beverage is usually stated as the percentage of alcohol by volume (ABV) or—in the United States—as proof. In the U.S.A., proof is twice the percentage of alcohol by volume at 60 degrees Fahrenheit (e.g., 80 proof = 40 per cent ABV). Degrees proof were formerly used in the U K, where 100 degrees proof was equivalent to 57.1 per cent ABV. Historically, this was the most dilute spirit that would sustain the combustion of gunpowder.

Ordinary distillation cannot produce alcohol of more than 95.6 per cent ABV (191.2 proof) because at that point alcohol is an azeotrope with water. Alcohol of this high level of purity is commonly called neutral grain spirit.

Most forth appearing in fermentation cannot reproduce when the concentration of alcohol is higher than about 18 per cent, so that is the practical limit for the strength of fermented beverages such as wine, beer, and sake. Strains of yeast have been developed that can reproduce in solutions of up to 25 per cent ABV.

~~SERVING SIZES~~

Beer Characteristics

The characteristics of beer are given below:-

The two major categories of beer are ales and lagers, and the forth appearing in fermentation used in chemical change by fermenting determines the differences between the two. Ales are fermented with Saccharomyces cerevisiae at warm temperatures for short periods of time, typically a week. This forth appearing in fermentation ferments in warmer temperatures to produce a fruity flavour that characterizes ale. Ales are aged, or conditioned, for one to three weeks. Ales should be served at room or cellar temperature—about 13°C.

Biggest are usually drier, crisper, and less fruity in taste than ales. Biggest are fermented with Saccharomyces uvarum at cooler heat of air for longer periods up to two weeks, which encourages the flavour components in lagers to combine and mellow. Long time aged for up to 12 weeks. Lagers are best served chilled.

All beer can be further described by three additional variables: colour, bitterness, and gravity. Colour is determined by the type and amount of malt used to make the beer. For e.g., lightly roasted malt produces a light-coloured beer.

Bitterness depends on the type and amount of grain and hops used in the beer's production. Beers made with wheat, for e.g., are generally less bitter than beers made with dark-roasted barley.

Gravity refers to the beer's density, or the amount of dissolved sugars in the wort. As more sugars dissolve in the brewing water during mashing, the density of the wort become greater. The forth appearing in fermentation consumes these sugars to produce alcohol during chemical change by fermenting, so elevated sugar levels translate into a higher alcohol content in the finished beer.

Organic acid

The relative stability of the conjugate base of the acid determines its acidity. Other groups can also confer acidity, usually weakly: -OH, -SH, enol group, and the phenol group. In biological systems organic compounds containing only these groups are not generally referred to as organic acids.

The organic acid is combination of organic compound and acidic function this is most organic acid which is associated the carboxylic acid and carboxyl group COOH. The H2So4 containing group S2oH.

LACTIC ACID

The lactic acid is called milk acid which is play the role in biochemical processes because lactic acids is a chemical compound. It was first isolated in 1780 by a Swedish chemist, Carl Wilhelm Scheele, and is a carboxylic acid with a chemical formula of C3H6O3. It has a hydroxyl group adjacent to the carboxyl group, making it an alpha hydroxy acid (AHA). In solution, it can lose a proton from the acidic group, producing the lactate ion CH3CH(OH)COO". It is miscible with water or ethanol, and is hygroscopic.

Lactic acid is chiral and has two optical isomers. One is known as L-(+)-lactic acid or (S)-lactic acid and the other, its mirror image, is D-(")-lactic acid or (R)-lactic acid. L-(+)-Lactic acid is the biologically a great consequence isomer.

Industrially, lactic acid chemical change by fermenting is performed by Lactobacillus bacteria, among others. These bacteria can operate in the mouth; the acid they produce is responsible for the tooth decay known as caries.

ACETIC ACID

Actic acid is organic acid and ethanoic acid which have sour test and pungent spet and the formula is CH3COOH. It is not strong acid but only partially dissociated acid. Pure, water-free acetic acid is a colourless liquid that absorbs water from the a surrounding object (hygroscopy), and freezes at 16.5 °C (62 °F) to a colourless crystalline solid. The pure acid, and concentrated solutions, are dangerously corrosive.

Actic acid is carboxylic acid and its momentous chemical reagent and companie chemical. That is used in the production of polyethylenc fereptithalase which is used soft drink as use in photographic filmand polyvinyl acetate for wood glue, as well as synthetic fibres and fabrics. In households, diluted acetic acid is often used in descaling agents. In the meal industry acetic acid is used under the meal additive code E260 as an acidity regulator.

The global demand of acetic acid is around 6.5 million tonnes per year of which approximately 1.5 Mt/a is met by recycling; the remainder is manufactured from petrochemical feedstocks or from biological sources. Dilute acetic acid produced by natural fermentation is called vinegar.

FORMIC ACID

The formic acid is the carboxylic acid and methanoic acid and formula is CHo2 (HCOOH). Basically this is most momentous in chemical synthesis and in venom of bee.

CITRIC ACID

The organic acid is a natural presercative which are used to soft drink and teste of meals we can say the critic acid have above all features because critic acid is a organic acid.

Citric acid exists in greater than trace amounts in a variety of fruits and vegetables, most notably citrus fruits. Lemons and limes have particularly high concentrations of the acid; it can constitute as much as 8 per cent of the dry weight of these fruits. The concentrations of citric acid in citrus fruits range from 0.005 mol/L for oranges and grapefruits to 0.30 mol/L in lemons and limes. These values vary depending on the cultivar and the circumstances in which the fruit was grown.

GLUCONIC ACID

Gluconic acid is organic compound and one of the 16 stereoisomers of 2,.....6 penthydroxyhexanoic acid and the formula is C6h12O7 and the structure formula is

CH2oH (CHoH)4 COOH
CHoH
CHoH
CHoH
CHoH
O - CHoH

Therapeutic compounds

A group of new compounds, N-(all-trans-Retinoyl)-L-cysteic acid, N-(13-cis-Retinoyl)-L-cysteic acid, N-(all-trans-Retinoyl)-L-cysteinesulfinic acid, N-(13-cis-Retinoyl)-L-cysteinesulfinic acid, N-(all-trans-Retinoyl)-L-homocysteic acid, N-(13-cis-Retinoyl)-L-homocysteic acid, and sodium salts of these compounds, adding sodium salts of their esters and amides, is shown to exhibit therapeutic effects per se, and which compounds in combination with cytotoxic compounds, such as docetaxel, paclitaxel, doxorubicin and mitoxantrone, exhibit a synergistic effect. These compounds make it possible to manufacture new formulations of poorly soluble pharmaceutical compounds, and the present invention discloses a process of manufacturing water-soluble formulations of such compounds, exemplified by docetaxel, and paclitaxel, exhibiting enhanced pharmacological activity, and formulations of water-soluble pharmaceuticals exemplified by doxorubicin and mitoxantrone, exhibiting improved therapeutic efficacy.

Antibiotic

Anantibiotic is a mixture which is used for kills the growth of bacteria. This is belong to broad a group of antimicrobial mixture.

The term "antibiotic" was coined by Selman Waksman in 1942 to describe any substance produced by a microorganism that is antagonistic to the growth of other microorganisms in high dilution. This original definition excluded naturally occurring substances that kill bacteria but are not produced by microorganisms and also excluded synthetic antibacterial compounds such as the sulfonamides.

With advances in medicinal chemistry, most antibiotics are now semisynthetic—modified chemically from original compounds found in nature, as is the case with beta-lactams. Some antibiotics are still produced and isolated from living organisms, such as the aminoglycosides, and others have been created through purely synthetic means: the sulfonamides, the quinolones, and the oxazolidinones. In addition to this origin-based classification into natural, semisynthetic, and synthetic.

Antibiotics can be classified into two parts according their effect on micro-organisms as first the kill of bacteria and impair bacterial growth are called bacteriostatic agents.

Bacteriocin

Bacteriocins are proteinaceous toxins produced by bacteria to inhibit the growth of similar or closely related bacterial strain(s). They are typically considered to be narrow spectrum antibiotics, though this has been debated They are phenomenologically analogous to preparation used for raising dough and paramecium killing factors, and are structurally, functionally, and ecologically diverse. Bacteriocins were first discovered by A. Gratia in 1925. He was involved in the process of searching for ways to kill bacteria.

Classification of Bacteriocins

Bacteriocins are classification in several ways like that. There are several large categories of bacteriocin which are only phenomenologically related. These include the bacteriocins from gram-positive bacteria, the colicins, the microcins, and the bacteriocins from Archaea. The bacteriocins from E. coli are called colicins (formerly called 'colicines,' meaning 'coli killers'). They are the longest studied bacteriocins. They are a diverse group of bacteriocins and do not include all the bacteriocins produced by E. coli. For example, the bacteriocins produced by Staphylococcus warneri are called as warnerin or warnericin. In fact, one of the oldest known so-called colicins was called colicin V and is now known as microcin.

This is very small which is secreted in categorised manner and classic clicins. The lactic acid fermenting.

(i) mechanism of killing
(ii) producing strain
(iii) common resistance mechanisms

Erythromycin

The erythromycin has an antimicrobial spectrum or pemicillin for use of allergy and it is also known macnolide antibiotic. For respiratory tract infections, it has better coverage of atypical organisms, adding mycoplasma and Legionellosis. It was first marketed by Eli Lilly and Company, and it is today commonly known as EES (erythromycin ethylsuccinate).

This macrocyclic compound contains a 14-membered lactone ring with ten asymmetric centres and two sugars (L-cladinose and D-desoamine), making it a compound very tipical to produce via synthetic methods. Erythromycin is generated from actinomcete saccharopolyspora erythraca.

Bread-Making

This article will show how simple it is to make a loaf of bread. Remember, forth appearing in fermentation bread making is easy to do. It is just a little time consuming. But, oh what a wonderful way to spend your time creating a warm, fragrant loaf of bread.

The below steps are a general way in which to make yeast bread. When making your bread, follow the specifics of the recipe, not just these steps.

Biogas

Biogass is refer to a gas generated by biological breakdown of organic matter in a space of oxygen we can say the biogass is a type of biofuel and it is originated from biogenic material.

One type of biogas is generated by anaerobic digestion or chemical change byfermenting of biodegradable materials such as biomass, manure, sewage, municipal waste, green waste and energy crops. This type of biogas comprises primarily methane and Co2. The other principal type of biogas is wood gas which is created by gasification of wood or other biomass. This type of biogas is comprised primarily of nitrogen, hydrogen, and carbon monoxide, with trace amounts of *methane.*

The gases methane, hydrogen and carbon monoxide can be combusted or oxidised with oxygen. Air contains 21 per-cent oxygen. This energy release allows biogas to be used as a fuel. Biogas can be used as a low-cost fuel in any country for any heating purpose, such as cooking. It can also be used in modern waste management facilities where it can be used to run any type of heat engine, to generate either mechanical or electrical power. Biogas can be compressed, much like natural gas, and used to power motor vehicles and in the UK, for example, is estimated to have the potential to replace around 17 per cent of vehicle fuel. Biogas is a renewable fuel, so it qualifies for renewable energy subsidies in some parts of the world.

Production

Biogass is generated as landfull gas or digester gas and biogas plant is given to anaerobic digenter from energy crops.

Biogas can be generated utilising anaerobic digesters. These plants can be fed with energy crops such as maize silage or biodegradable wastes adding sewage sludge and meal waste. During the process, an air-tight tank transforms biomass waste into methane producing renewable energy that can be used for heating, electricity, the different-2 other terms that use any variation of an internal combustion engine. There are two key processes: Mesophilic and Thermophilic digestion.

Landfill gas is generated by wet organic waste decomposing under anaerobic conditions in a landfill. The waste is covered and mechanically compressed by the weight of the material that is deposited from above. This material prevents oxygen exposure thus allowing anaerobic microbes thrive. This gas builds-up and is slowly released into the atmosphere if the landfill site has not been engineered to capture the gas. Landfill gas is hazardous for three key reasons. Landfill gas becomes explosive when it escapes from the landfill and mixes with oxygen. The lower explosive limit is 5 percent methane and the upper explosive limit is 15 per cent methane. The methane contained within biogas is 20 times more potent as a greenhouse gas than carbon dioxide. So uncontained landfill gas which escapes into the atmosphere may significantly contribute to the effects of global warming. In addition landfill gas' impact in global warming, volatile organic mixture (VOCs) contained within landfill gas contribute to the formation of photochemical smog.

Composition

Biogas is depend on the origin of anaerobic digention process because landfill gas has concentrations 50 per cent waste manner of appling remedies technologies con and also contain, tractional water vapor, function of biogas heat of air.

Deenabandhu Model

The Deenabandhu Model is a new biogas-generatd model popular in India. The unit usually has a capacity of 2 to 3 cubic metres. It is constructed using bricks or by a ferrocement mixture. The brick model costs approximately 18,000 rupees and the ferrocment model 14,000 rupees, above all India's Ministry of Non-conventional Energy Sources offers a subsidy of up to 3,500 rupees per model constructed.

ARTI Biogas Plant

ARTI biogas developed acompact biogas plant which is used meal waste the plant is compact to be used both urban and rural households in India.

Few ARTI biogas plants have been installed in other parts of India or the world. The design and development of the ARTI biogas plant won the 2006 Ashden Award for Sustainable Energy 2006 in the meal Security category.

Gober Gas

Gober gas is biogas generated from cow dung. A gober gas plant is an airtight circular pit made of concrete with a pipe connection. First, manure is dumped in the pit. Then, water or wastewater is included to the manure and the concoction is sealed in the airtight concrete pit with a gas pipe leading to stove unit in the kitchen serving as the only egress for gas. When the control valve on the gas pipe is opened the biogas is combusted for cooking in a largely odourless and smokeless manner. After the anaerobic microbial process has been exhausted, the residue left in the concrete pit is often used as fertilizer. Owing to the process's simplicity in implementation and use of cheap raw materials, it is often regarded as one of the most environmentally sound energy sources for rural needs.

Vitamin

A vitamin is an organic mixture required as a nutrient in tiny amounts by an organism. The term 'vitamin' first became popular in the early-1800's as a contraction of the words 'vital' and 'mineral', though the actual meaning of the word has developed somewhat since that time. A mixture is called a vitamin when it cannot be synthesised in sufficient quantities by an organism, and must be obtained from the diet. Thus, the term is conditional both on the circumstances and the particular organism. For e.g., ascorbic acid functions as vitamin C for some animals but not others, and vitamins D and K are required in the human diet only in certain circumstances. The term vitamin does not add other essential nutrients such as dietary minerals, essential fatty acids, or essential amino acids, nor does it encompass the large number of other nutrients that promote health but are otherwise required less often.

Vitamins are categorised by their biological and chemical activity. Thus, each "vitamin" may refer to several vitamer compounds that all show the biological activity associated with a particular vitamin. Such a set of chemicals are grouped under an alphabetised vitamin "generic descriptor" title, such as "vitamin A", which add the compounds retinal, retinol, and many carotenoids. Vitamers are often inter-converted in the body.

Vitamins have diverse biochemical functions, adding function as hormones (e.g., vitamin D), antioxidants (vitamin E), and mediators of cell signaling and regulators of cell and tissue growth and differentiation (vitamin A). The largest number of vitamins (B complex vitamins) function as precursors for enzyme cofactor bio-molecules (coenzymes), that help act as catalysts and substrates in metabolism. When acting as part of a catalyst, vitamins are bound to enzymes and are called prosthetic groups. For e.g., biotin is part of enzymes involved in making fatty acids.

Vitamins also act as coenzymes to carry chemical groups between enzymes. For e.g., folic acid carries various forms of carbon group – methyl, formyl and methylene— in the cell. Although these roles in assisting enzyme reactions are vitamins' best-known function, the other vitamin functions are equally important.

Until the 1900s, vitamins were obtained solely through meal intake, and changes in diet can alter the types and amounts of vitamins ingested. Vitamins have been produced as commodity chemicals and made widely available as inexpensive pills for several decades, allowing supplementation of the dietary intake.

Riboflavin

The Vitamin B12 is also called riboflavin. It is absorbed micronitrient in maintaining health human. It is the central component of the cofactors FAD and FMN, and is therefore required by all flavoproteins. As such, vitamin B2 is required for a wide variety of cellular processes. Like the other B vitamins, it plays a key role in energy metabolism, and is required for the metabolism of fats, ketone bodies, carbohydrates, and proteins.

Milk, cheese, leafy green vegetables, liver, kidneys, legumes such as mature soybeans, yeast, mushrooms and almonds are good sources of vitamin B2, but exposure to light destroys riboflavin.

The name "riboflavin" comes from "ribose" and "flavin".

Cyanocobalamin

Cyanocobalamin is belong to common vitamin of tamer of vitamin B12 family, because this is most popular of vitamer of family because it is chemically the most air-stable, and it is the easiest to crystallise and so easiest to purify after it is generated by bacterial chemical change by fermenting. The cyanide is added to the molecule by activated charcoal columns in purification. Thus, the use of this form of B12 is the most wide-spread.

The correct chemical name of vitamin B12 actually is cyanocobalamin. In fact, vitamin B12 is the name for a whole class of chemicals with B12 activity, and cyanocobalamin is only one of these. Cyanocobalamin usually does not even occur in nature, and is not one of the forms of the vitamin which is directly used in the human body. However, animals and humans can convert it to active (cofactor) forms of the vitamin, such as methylcobalamin.

The act of producing of vitamin B12

The fascinating molecules in the world of science and medicine is vitamin B12 (cobalamin), which was originally discovered as the anti pernicious anemia factor and whose enigmatic complex structure is matched only by the beguiling chemistry that it mediates. The biosynthesis of this essential nutrient is intricate, involved and, remarkably, confined to certain members of the prokaryotic world, seemingly never have to have made the eukaryotic transition. In humans, the vitamin is required in trace amounts to assist the actions of only two enzymes, methionine synthase and (R)-methylmalonyl-CoA mutase; yet commercially more than 10 t of B12 are produced each year from a number of bacterial species. The rich scientific history of vitamin B12 research, its biological functions and the pathways employed by bacteria for its de novo synthesis are described. Current strategies for the improvement of vitamin B12 production using modern biotechnological techniques are outlined.

Industrial of Microbial Enzymes

The industrial microbiology, is concerned with all aspects of business that relate to microbiology. In a more restricted sense, industrial microbiology is concerned with

(i) employing microorganisms to produce a desired product, (ii) preventing microbes from diminishing the economic value of various products. This dual purpose is clearly seen in the meal.

Various commercial products of economic value made by microbes are.

1. medicines, i.e., pharmaceuticals, including antibiotics, steroids, human protein, vaccines, and vitamins,
2. organic solvents]
3. amino acids,
4. enzymes,
5. recovery of metals from ores through bioleaching,
6. organic acids,
7. synthetic fuels. In addition to these, quite recently potential of microbes could also be realised in,
8. alcohols,
9. recovery of petrol,
10. single all protein production.

The term chemical change by fermenting in industrial microbiology is used in a wider sense to add any chemical transformation of organic mixture carried out by using microbes and their enzymes.

Production methods in industrial microbiology bring together the raw materials (substrates), microorganisms (specific strains or microbial enzymes) and a controlled favourable a surrounding object to produce the desired substance.

Industrial microbiology is an momentous area of applied microbiology. It refers to the use of microorganisms in commercial enterprise and. Cheap raw materials are converted to valuable products through the metabolism of microbes. Microbes for this purpose could be exploited in different ways.

(i) synthesis of fermentation products as acids, alcohols or other organic compounds
(ii) transformation of one mixture into another desired type,
(iii) The act of producing of enzymes, antibiotics, or insecticides, or
(iv) the use of microbes themselves as meal.

Beer Fermentation

Fermentation is the process by which fermentable carbohydrates are converted by forth appearing in fermentation into alcohol, (Co2) and numerous other byproducts. It is these byproducts that have a considerable effect on the taste, aroma, and other properties that characterize the style of beer.

Pitching Beer Yeast

Strain

The yeast strain itself is a major contributor to the flavour and character of the beer. Thus the choice of yeast strain depends on such things as the oxygen requirements, cropping methods, attenuation limits, fermentation rate, fermentation temperatures, flocculation characteristics, and the flavour profile but esters, fusel alcohols sulfur.

Pitching Rates

The brewer's ability to pitch the correct number of forth appearing in fermentation cells to initiate fermentation is crucial to consistently producing a product of superior and constant quality. Pitching rates are governed by a number of factors, adding yeast strain, fermentation capacity of the yeast, yeast viability, flocculation characteristics, previous history of the yeast, and desired beer flavour characteristics. Other considerations when choosing pitching rates add wort gravity, wort constituents, fermentation temperature, and the degree of wort aeration. For e.g., highly flocculent yeast strains may settle prematurely, requiring the brewer to either over-pitch or to mix and aerate the yeast by "rousing."

Methods for Determining Pitching Rates

Traditionally, brewers determined pitching rates on the basis of either the volume or the weight of yeast slurry to a known quantity of wort. Today most breweries determine pitching rates by a number of alternative methods adding the use of a counting chamber (hemacytometre), centrifugation of the forth appearing in fermentation slurry, electron cell counting, or of in-line biomass sensors.

Microbiological Contamination

Uninoculated wort is liable to contamination by many types of mold; bacteria such as *Pediococcus*, spp. and *Lactobacillus* spp.; and wild yeasts such as *Hansenula*, *Dekkera*, *Brettanomyces*, *Candida*, and *Pichia*. Other *Saccharomyces* species may be present.

LAGER FERMENTATIONS

Traditional European lager brewers, especially in Germany and Scandinavia, use starter tanks in which yeast is included to the cooled wort and the contents then allowed to stand for a period of up to 24 hours before racking. Traditionally, starter tanks were open vessels, but nowadays they are generally designed like regular closed-in fermenters.

Topping Up (Darauflassen)

If the brewer doesn't have enough forth appearing in fermentation or if the brewer has a smaller mash tun and kettle than the fermenter, the brewer can employ a technique called topping up or darauflassen. Topping up, a technique common among German lager brewers, is the infusion of wort into a tank with strongly fermenting young beer.

Traditional Lager Fermentation

While vertical cylindroconical fermenters are used to make the majority of the world's lager production, open square fermenters are still commonly used (Eastern Europe) in traditional lager fermentation. In some major brewers have started fermenting their lagers in open square fermenters for quality reasons.

Modern Lager Fermentation

Today modern lager fermentation typically uses vertical cylindrodroconical fermenters achieving similar flavour profiles compared to traditional fermentation systems. In modern fermentation systems, the yeast is pitched at higher temperatures between 7 and 8°C, and after a couple of days the temperature is increased to 10 to 11°C (25). After 3 to 4 days, at peak fermentation, the chemical change by feremtomg temperature is allowed to ramp up in order to facilitate a rapid reduction in diacetyl. Some brewers use the same starting temperature for pitching but then become greater the temperature between 14 and 15°C (24). Other breweries are known to pitch between 10 and 13°C and then become greater the temperature to as high as 17°C, with a short diacetyl rest (28).

ALE FERMENTATIONS

Starter Tanks

Like lager brewers, some ale brewers initiate chemical change by fermenting in starter tanks in which the residence time may be as short as 3 hours or as long as 36 hours.

Traditional Ale Fermentation

The brewing of traditional cask ale is fermented in shallow vessels that could be round, square or rectangular often referred by the type of fermentation system (Yorkshire square, the Burton Union, the ale top-skimming system, and the ale dropping system). The Burton Union system is now confined to a single brewery in Burton-on-Trent, and the Yorkshire square system is found in Yorkshire and Midland breweries.

Fermentation Systems

Traditional Lager Fermentation System

The traditional method of collecting lager forth appearing in fermentation is to decant the "green" beer from the settled forth appearing in fermentation on the bottom of the open fermenters. The yeast is collected manually from the middle layer of the sediment on the floor.

Traditional Ale Fermentation System

Traditionally, top-cropped ale yeast was harvested by skimming the head of forth appearing in fermentation that formed on top of the beer in the shallow, flat-bottomed chemical change by fermenting vessel. Generally, the second crop that forms towards the end of fermentation is the one that is harvested since the yeast is pure, with very high viability.

Cylindroconical Fermenters

Today, with the advent of cylindroconical fermenters, the differentiation on the basis of bottom and top cropping has become less distinct. Cylindroconical tanks allow improved forth appearing in fermentation separation and collection strategies for both lager and ale forth appearing in fermentation. The angle at the bottom of the tank allows the forth appearing in fermentation to settle into the base of the vessel at the completion of primary fermentation. This aids in yeast collection from the bottom of the cone without exposing the yeast to outside air, leaving the beer comparatively free of yeast.

Traditional Ale Top-Skimming System

This is the traditional ale skimming system used in the United Kingdom. It utilizes top-fermenting forth appearing in fermentation cropped for re-pitching by skimming from the surface. The vessels are normally shallow and flat-bottomed and may be round, square, or rectangular. Traditionally, these vessels are open-topped to facilitate skimming and are located in a well-ventilated room to disperse the carbon dioxide evolved during chemical change by fermenting.

Traditional Ale Dropping System

This is a variation of the ale top-skimming system, whereby the wort is pitched, collected, and partially fermented in one vessel (the settling tanks) and then dropped or transferred to a second vessel after 24 to 36 hours to complete the fermentation (53). Trub and cold break are left behind in the first vessel, and the transfer also rouses the fermentation.

Yorkshire Square System

This system is used in the north of England to produce ales with a clean, round palate. It is a variation of the ale chemical change by fermenting system adapted to enable the use of a flocculent forth appearing in fermentation strain or a two-strain system with one of the yeasts being flocculent. The vessels are normally square or rectangular in shape, and constructed of stainless steel. The Yorkshire square system consists of a lower compartment with a gently sloping upper deck located just above the fill level of the vessel.

Burton Union System

This system was commonly used in Burton-on-Trent and produces pale ales with a fruity character by using "powdery" non-flocculent forth appearing in fermentation. The wort is collected and pitched in a separate vessel and then transferred to the union set after 24 to 48 hours. The union set consists typically of 2 rows of wooden casks, 24 to 50 in number and each of 7 hl capacity, positioned under a cooled trough. Each cask is fitted with a swan neck that overhangs a slightly inclined trough mounted above the casks.

Open Square Fermenters

Traditional lager systems in Continental Europe and still common in Eastern Europe utilize open square fermenters similar to those used in the traditional ale top system.

Dual Purpose Vessel System

The vessel characteristics that make cylindroconicals so suitable for fermenting vessels also make them ideal for conditioning tanks. This has led to the installation of dual-purpose vessels where primary chemical change by fermenting and conditioning are carried out in the same vessel. Dual-purpose vessels have cooling jackets located high in the vessel for chemical change by fermenting and low in the vessel for conditioning. Other differences are that conditioning tanks are normally required to be top-pressured to maintain carbonation levels, which is not required in fermenters.

Fermentation (wine)

The process of chemical change by fermenting in wine is the catalyst function that turns grape juice into an alcoholic beverage. During chemical change by fermenting yeast interact with sugars in the juice to create ethanol, commonly known as ethyl alcohol, and carbon dioxide (as a by-product). In winemaking the temperature and speed of chemical change by fermenting is an momentous consideration as well as the levels of oxygen present in the must at the start of the chemical change by fermenting. The risk of stuck chemical change by fermenting and the development of several wine faults can also occur during this stage which can last anywhere from 5 to 14 days for primary chemical change by fermenting and potentially another 5 to 10 days for a secondary chemical change by fermenting. Chemical change by fermenting may be done in stainless steel tanks, which is common with many white wines like Riesling, in an open wooden vat, inside a wine barrel and inside the wine bottle itself as in the production of many sparkling wines.

HIGH GRAVITY FERMENTATIONS

High-gravity chemical change by fermenting involve worts of up to 18°P and even higher. Following chemical change by fermenting and maturation, the beer is diluted with cool carbonated water to a prescribed original gravity or to a prescribed alcohol concentration. There are a number of feature associated with high-gravity brewing. Munroe reports that it results in beers that are more consistent (per cent alcohol, original gravity, etc.) and more physically stable since the mixtures responsible for haze are more easily precipitated at higher concentrations. Handling more concentrated wort results in increased utilization of equipment and lower energy costs. According to Munroe, the disadvantages are longer fermentation times, different flavour characteristics, and poorer hop utilization than normal gravity fermentations.

Malolactic Fermentation

Instead of yeast, bacteria plays a fundamental role in malolactic chemical change by fermenting which is essentially the conversion of malic acid into lactic acid. This has the benefit of reducing some of the tartness and making the resulting wine taste softer. Depending on the style of wine that the winemaker is trying to produce, malolactic chemical change by fermenting may take place at the same time as the yeast chemical change by fermenting.

Because yeasts perform this process in the absence of oxygen, ethanol chemical change by fermenting is classified as anaerobic.

Ethanol chemical change by fermenting occurs in the production of alcoholic wine, beer and other liquid and ethanol fuel, and in the rising of bread dough.

THE CHEMICAL PROCESS OF CHEMICAL CHANGE BY FERMENTING OF GLUCOSE

The chemical equation below summarizes the fermentation of glucose. One glucose molecule is converted into 2C2HOH and two C6H12oO6 molecules:

C6H12O6 '! 2C2H5OH + 2CO2

The process begins with a molecule of glucose being broken down by the process of glycolysis into pyruvate:

C6H12O6 '! 2 CH3COCOO" + 2H+

This reaction is accompanied by the size difference of two molecules of NAD+ to NADH and a net of two ADP molecules converted to two ATP plus the two water molecules.

Pyruvate is then converted to acetaldehyde and Co2 by an enzyme called pyruvate decarboxylase and requiring thiamine diphosphate as cofactor. The acetaldehyde is subsequently reduced to ethanol by the NADH from the previous glycolysis, which is returned to NAD+:

CH3COCOO" + H+ '! CH3CHO + CO2

CH3CHO + NADH ’! C2H5OH + NAD+

Many species of forth appearing in fermentation will oxidize pyruvate completely to carbon dioxide and water (respiration) if oxygen is present in the surrounding object and will ferment only in an anaerobic surrounding object. Above all the commonly used bakers’ yeast S. cerevisiae as well the yeast S. pombe, both prefer fermentation to respiration even in the presence of oxygen and will yield ethanol even under aerobic conditions given the right sources of nutrition.

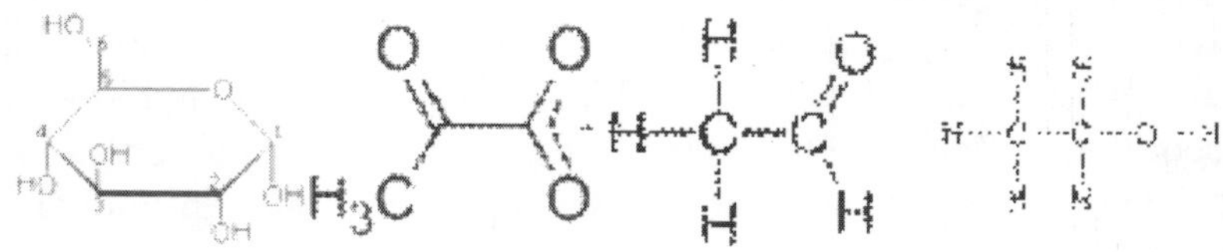

Glucose Pyruvate Acetaldehyde (CH3CHO) Ethanol (C2H5OH)

Procedure of making alcohol from sugar:

USES

Ethanol fermentation is responsible for the rising of bread dough. Forth appearing in fementaion organisms consume sugars in the dough and produce ethanol and Co2 as waste products. The Co2 forms bubbles in the dough, expanding it into something of a foam. Nearly all the ethanol evaporates from the dough when the bread is baked.

The production of all alcoholic beverages, except those produced by carbonic maceration, employs ethanol fermentation by forth appearing in fermentation. Wines and brandies are produced by chemical change by fermenting of the natural sugars present in fruits, especially grapes. Beers, ales, and whiskeys employ fermentation of grain starches that have been converted to sugar by the application of the enzyme, amylase, which is present in grain kernels that have been germinated. Amylase-treated grain or amylase-treated potatoes are fermented for the production of vodka. Chemical change by fermenting of cane sugar is the first step in producing rum. In all cases, the chemical change by fermenting must take place in a vessel that is arranged to allow Co2 to escape, but that prevents outside air from coming in, as exposure to oxygen would prevent the formation of C2H5OH.

Similar yeast fermentation of various carbohydrate products is used to produce much of the ethanol used for fuel.

USES

Ethanol fermentation is responsible for the rising of bread dough. Yeast organisms consume sugars in the dough and produce ethanol and carbon dioxide as waste products. The carbon dioxide forms bubbles in the dough, expanding it into something of a foam. Nearly all the ethanol evaporates from the dough when the bread is baked.

The production of all alcoholic beverages, except those produced by carbonic maceration, employs ethanol fermentation by yeast. Wines and brandies are produced by fermentation of the natural sugars present in fruits, especially grapes. Beers, ales, and whiskeys employ fermentation of grain starches that have been converted to sugar by the application of the enzyme, amylase, which is present in grain kernels that have been germinated. Amylase-treated grain or amylase-treated potatoes are fermented for the production of vodka. Fermentation of cane sugar is the first step in producing rum. In all cases, the fermentation must take place in a vessel that is arranged to allow carbon dioxide to escape, but that prevents outside air from coming in, as exposure to oxygen would prevent the formation of ethanol.

Similar yeast fermentation of various carbohydrate products is used to produce much of the ethanol used for fuel.

FEEDSTOCKS FOR FUEL PRODUCTION

The dominant ethanol feedstock in warmer regions is sugarcane. In temperate regions, this accessibility has been somewhat replicated by selective breeding of the sugar beet.

In the United States, the main feedstock for the production of ethanol is currently corn. Approximately 2.8 gallons of ethanol are produced from one bushel of corn (0.42 litre per kilogram). While much of the corn turns into ethanol, some of the corn also forth appearing in fermentation by-products such as DDGS distillers dried grains with syllables that can be used to fulfill a portion of the diet of livestock. A bushel of corn produces about 18 pounds of DDGS.. Although most of the fermentation plants have been built in corn-producing regions, sorghum is also an momentous feedstock for ethanol production in the Plains states. Pearl millet is showing promise as an ethanol feedstock for the southeastern U.S., and duckweed potential is being studied.

In some parts of Europe, particularly France and Italy, wine is used as a feedstock due to a massive oversupply termed wine lake. Japan is hoping to use rice wine as an ethanol source.

6

SELECTION, SCALE UP AND SCALE DOWN

The fermentation technology research is 'Scaling up' and 'Scaling down' studies. However, the phrase 'scaling up' is more commonly understood and practiced during the designing of industrial scale fermentors. Where as 'scaling down' studies are rarely heard that frequent many fermentation technologists are not aware that during most times of their work they are doing 'scale down studies' Maybe the phrase 'scale up' has more impact factor than 'scale down' studies. For example

When we intend to start with a chemical change by fermenting process with the ultimate objective of producing the chemical change by fermenting products on the level of industrial scale. Products need to be produced at large volume so that the process is economically viable. This requires scaling up. We do scaling up studies to ensure that the fermentation process is technically and economically viable to be produced in the end at a large scale.

SCALE UP STUDIES

Scale up studies are studies carried out at the laboratory or even pilot plant scale fermentors to give in return for data that could be used to extrapolate and build the large scale industrial fermentors with sufficient confidence it will function properly with all its behaviours anticipated. More momentous during scale up exercises you are trying to build industrial size fermentor capable or close of producing the fermentation products as efficient as those produced in small scale fermentors.

IF SCALE UP STUDIES FAILED?

The chemical change by fermenting process is technically not possible and is a failure than we have two options:

1. Either find the cause of the failure or back to the drawing board!
2. Abort the whole project with minimum economic loss to the investors

The investors and engineers need more confidence in predicting how the behaviour of the chemical change by fermenting will occur at the level of industrial scale. If it fails it might mean millions of dollars lost in the exercise where failure was not seen. So scaling up will give the investors more confidence on the chances of success and against economic disasters.

The exercise in scaling up involved a number of programmed research or steps that has to be established so as to predict the final behaviour of the final large scale production fermentor. This process are include:-

1. Inoculum development
2. Sterilization establishing the correct sterilisation cycle at larger loads
3. Environmental parametres such as nutrient availability, ph, temperature, dissolved. oxygen, dissolved carbon dioxide,
4. Shear conditions, foam production

SCALE DOWN STUDIES

The main objective is to carry out studies on smaller bioreactors in order to gain data and confidence and predict the behaviour how things actually will behave in large production fermentor.

Scale down studies are also used while during the operation of large industrial scale fermentors in trouble shooting or trying to optimize the industrial scale chemical change by fermenting. This method is called the fermentation monitoring.

The goal when scaling down is to create a small-scale or lab-scale system that mimics the performance of its large-scale (pilot or manufacturing) counterpart, when both the process parametres are varied within their operating ranges and also when a process parametre deviates outside its operating range.

The main type of studies in scale down such as:

1. Medium design
2. Medium sterilisation
3. Inoculation procedures
4. Number of generations
5. Mixing
6. Oxygen transfer rate

RULES OF SCALE DOWN

In scale down a few rules that should be followed:

SIMILAR FERMENTOR GEOMETRY

1 Similar model geometries and ratios of system. The impeller and sparger designs, and placements within the vessel must be identical or similar. Wrong models used in scale down studies might invalidate the data obtained. Since a typical fermentation process might involved different fermentor capacities, scaling down will therefore be very challenging and proper strategies need to be developed during scale down studies.

SIMILAR METHODS OF MEASURING ANALYSING SAMPLES

Similar methods of analyses and monitoring be applied at scale down studies such as:

1. sample-dilution schemes and measurement times for calculating culture optical densities,
2. wet and dry cell-weights,
3. media metabolite levels.

SAMPLING VOLUME

Agitation Rate

The scale down studies to provide either representative:

1. oxygen transfer rate,
2. tip speed,
3. Reynolds Number, or
4. power-input per unit volume,

CULTURE GROWTH

Product Yield,
Process sensitivity and
Do scale down studies a few time.

Inoculum Development

The preparation of a population of micro-organisms from a dormant stock culture to an active state of growth that is suitable for inoculation in the final production stage is called inoculum development. As a first step in inoculum development, inoculum is taken from a working stock culture to initiate growth in a suitable liquid medium.

Bacterial vegetative cells and spores are suspended, usually, in sterile tap water, which is then included to the broth. In case of non-sporulating fungi and actinomycetes the hyphae are fragmented and then transferred to the broth.

Inoculum development is generally done using flask cultures; flasks of 50 ml to 12 litres may be used and their number can be advanced as per need. Where needed, small fermenters may be used.

Inoculum development is usually done in a stepwise sequence to grow the volume to the desired level. At each step, inoculum is used at 0.5-5 per cent of the medium volume; this allows a 20-200-fold increase in inoculum volume at each step. Typically, the inoculum used for production stage is about 5 per cent of the medium volume.

Sterilisation

Sterilisation refers to any process that effectively kills or eliminates transmissible agents from a surface, equipment, article of meal or medication, or biological culture medium. Sterilisation does not, above all remove prions. The transmissible agents is fungi bacteria viruses, spore forms. Sterilization can be achieved through application of heat, chemicals, irradiation, high pressure or filtration.

HEAT STERILISATION

Steam Sterilization Utensils

A widely-used method for heat sterilisation is the autoclave, sometimes called a converter. Autoclaves commonly use steam heated to 121–134 °C (250–273 °F). To achieve sterility, a holding time of at least 15 minutes at 121 °C (250 °F) or 3 minutes at 134 °C (273 °F) is required. Additional sterilising time is usually required for liquids and instruments packed in layers of cloth, as they may take longer to reach the required temperature (unnecessary in machines that grind the contents prior to sterilisation). Following sterilisation, liquids in a pressurised autoclave must be cooled slowly to avoid boiling over when the pressure is released. Modern converters operate around this problem by gradually depressing the sterilisation chamber and allowing liquids to evaporate under a negative pressure, while cooling the contents.

Proper autoclave manner of appling remedies will inactivate all fungi, bacteria, viruses and also bacterial spores, which can be quite resistant. It will not necessarily eliminate all prions.

For prion elimination, various recommendations state 121–132 °C (250–270 °F) for 60 minutes or 134 °C (273 °F) for at least 18 minutes.

CHEMICAL STERILISATION

Chemicals are also used for sterilisation. Although heating provides the most reliable way to rid objects of all transmissible agents, it is not always appropriate, because it will damage heat-sensitive materials such as biological materials, fibre optics, electronics, and many plastics. Low temperature gas sterilisers function by exposing the articles to be sterilised to high concentrations of very reactive gases (and oxidising agents such as hydrogen peroxide and ozone). Liquid sterilants and high disinfectants typically add oxidising agents such as hydrogen peroxide and peracetic acid and aldehydes such as glutaraldehyde and more recently o-phthalaldehyde. While the use of gas and liquid chemical sterilants/high level disinfectants avoids the problem of heat damage, users must ensure that article to be sterilised is chemically compatible with the sterilant being used. The make the product by machine and other a source of the article can provide specific information regarding compatible sterilants. In addition, the use of chemical sterilants poses new challenges for workplace safety. The chemicals used as sterilants are designed to destroy a wide range of pathogens and typically the same properties that make them good sterilants makes them harmful to humans. Employers have a duty to ensure a safe work a surrounding object and work practices, engineering controls and monitoring should be employed appropriately. The occupational safety ad health act. 1970, sections for U. S.

Ozone

Ozone is used in industrial settings to sterilise water and air, as well as a disinfectant for surfaces. It has the benefit of being able to oxidise most organic matter. On the other hand, it is a toxic and unstable gas that must be produced on-site, so it is not practical to use in many settings.

Ozone offers many advantages as a sterilant gas; ozone is a very efficient sterilant because of its strong oxidising properties (E = 2.076 vs SHE, CRC Handbook of Chemistry and Physics, 76th Ed, 1995-1996) capable of destroying a wide range of pathogens, adding prions without the need for handling hazardous chemicals since the ozone is generated within the sterilizer from medical grade oxygen. In 2005 a Canadian company called TSO3 Inc received FDA clearance to sell an ozone steriliser for use in health care. The high reactivity of ozone means that waste ozone can be destroyed by passing over a simple catalyst that reverts it back to oxygen and also means that the cycle time is relatively short (about 4.5 hours for TSO3's model 125L). The downside of using ozone is that the gas is very reactive and very hazardous. The NIOSH immediately dangerous to life and health limit for ozone is 5 ppm, much 160 times smaller than the 800 ppm IDLH for ethylene oxide. Documentation for Immediately Dangerous to Life or Health Concentrations (IDLH): NIOSH Chemical Listing and Documentation of Revised IDLH Values (3/1/95) and OSHA has set the PEL for ozone at 0.1 ppm calculated as an eight hour time weighted average (29 CFR 1910.1000, Table Z-1). The Canadian Centre for Occupation Health and Safety provides an excellent summary of the health effects of exposure to ozone.

RADIATION STERILISATION

Methods of sterilisation exist using radiation such as electron beams, X-rays, gamma rays, or subatomic particles.

(1) Gamma rays are very penetrating and are commonly used for sterilization of disposable medical equipment, such as syringes, needles, cannulas and IV sets. Gamma radiation requires bulky shielding for the safety of the operators; they also require storage of a radioisotope (usually Cobalt-60), which continuously emits gamma rays (it cannot be turned off, and so always presents a hazard in the area of the facility).

(2) Electron beam processing is also commonly used for medical device sterilisation. Electron beams use an on-off technology and provide a much higher

dosing rate than gamma or x-rays. Due to the higher dose rate, less exposure time is needed and thereby any potential degradation to polymers is reduced. A limitation is that electron beams are less penetrating than either gamma or x-rays.

(3) X-rays, High-energy X-rays (bremsstrahlung) are a form of ionizing energy allowing to irradiate large packages and pallet loads of medical devices. Their penetration is sufficient to treat multiple pallet loads of low-density packages with very good dose uniformity ratios. X-ray sterilisation is an electricity based process not requiring chemical nor radioactive material. High energy and high power X-rays are generated by an X-ray machine that can be turned off for servicing and when not in use.

(4) Ultraviolet light irradiation is useful only for sterilisation of surfaces and some transparent objects. Many objects that are transparent to visible light absorb UV. UV irradiation is routinely used to sterilise the interiors of biological safety cabinets between uses, but is ineffective in shaded areas, adding areas under dirt (which may become polymerised after prolonged irradiation, so that it is very difficult to remove). It also damages many plastics, such as polystyrene foam.

(5) Subatomic particles may be more or less penetrating, and may be generated by a radioisotope or a device, depending upon the type of particle.

Irradiation with X-rays or gamma rays does not make materials radioactive. Irradiation with particles may make materials radioactive, depending upon the type of particles and their energy, and the type of target material: neutrons and very high-energy particles can make materials radioactive, but have good penetration, whereas lower energy particles cannot make materials radioactive, but have poorer penetration.

Environmental Parametres

The information on environmental variables that affect the living organisms is vital. Hydrographical studies play an momentous role in understanding the various biological processes (growth, physiology, reproduction.) and the general productivity of aquatic ecosystem.

Temperature and light play an momentous role in the gonadal gametogenesis, spawning and initiation of gonadal development. It is believed that the morphology of fish is determined by genetic and environmental factors, with these being particularly momentous during the early developmental stages when individuals are influenced by the surrounding object.

Estuaries may have the highest economic values per hectare of many aquatic surrounding object. Estuarine ecosystem is characterised by a variable salinity and temperature range of greater than Sea. In tropical estuaries seasonal changes in salinity is obviously greater than in temperature. Tropical and subtropical estuaries represent one of the most exploited ecosystems in the world. In India the estuaries vary from near 30-37 psu in salinity of the dry season. The Mandovi — Zuari estuarine systems are tropical estuarine systems and play major role in the economy of Goa.

Some of the early hydrographical and eco-biological work carried out on Goa estuaries are those of Dehadrai (1970); Cherian *et al.* (1975); Gangadhara Rao *et al.* (1976); Rao (1977); De Souza (1977); Qasim and Sen Gupta (1981); De Souza and Sen Gupta (1986); Ansari (1988); Padmavati and Goswami (1996); Ingole and Parulekar (1998). In the view of the importance of environmental condition and hydrographic parametres in fishery biology, it was felt necessary to study the environmental parametres such as temperature, salinity, dissolved oxygen and pH of the study area (Dona Paula in Zuari estuary).

Solid State Fermentation

In such chemical change by fermenting, microbial growth and product formation occur at the surface of solid substrates. Examples of such chemical change by fermenting are mushroom cultivation, mold ripened cheeses, starter cultures, etc. More recently, this approach has been used for the production of extracellular enzymes, certain valuable chemicals, fungal toxins, and fungal spores (biotransformation).

Traditional substrates are several agricultural products, rice, wheat, maize, soybean, etc. The substrate provides a rich and complex source of nutrients, which mayor may not need to be supplemented.

Such substrates selectively support mycelial organisms, which can grow at high nutrient concentrations and produce a variety of extracellular enzymes, e.g., a large number of filamentous fungi, and a few bacteria (Actinomycetes and one strain of Bacillus).

According to the physical state, solid state fermentations are divided into two groups:

(a) low moisture solids fermented without or with occasional/continuous agitation, and

(b) suspended solids fermented in packed columns, through which liquid is circulated. The fungi used for solid state chemical change by fermenting are usually obligate aerobes.

Solid state chemical change by fermenting on large scale use stationary or rotary trays. Temperature and humidity controlled air is circulated through the stacked solids. Less frequently, rotory drum type fermenters have been used.

Solid state chemical change by fermenting offer certain unique advantages, but suffer from some important disadvantages. Above all, commercial application of this process for biochemical production is chiefly confined to Japan.

Inoculum Preparation

Gas production during microbial chemical change by fermenting may be a useful marker for estimating the rate and extent of digestion of feeds. Measuring gas production instead of the disappearance of substrates such as fibre or dry matter has several advantages. It measures the sum total of all chemical change by fermenting and estimates the total energy contribution of the rumen to the cow from fibre, soluble carbohydrates and proteins. It is not contaminated by microbial residues like residual dry matter, which complicates the use of dry matter residues to estimated digestion kinetics. Gas production is much easier to measure than fibre disappearance because serial samples do not have to be analysed individually.

Recent work indicated that the source of inoculum had a significant impact on in vitro results. Preliminary experiments suggested that the method for preparing the inoculum also affected results. Initally, inoculum was prepared by blending ruminal solids with chilled in vitro media to detach microbes that cling to particles. Strained ruminal fluid was combined with the strained blended media to produce the inoculum. Above all, this inoculum yielded relatively high gas production for blanks (in which no sample material is included), and cooling the inoculum probably slowed initial chemical change by fermenting rates. The objectives of these studies were to determine if removing solids by centrifugation would reduce blank gas production and affect gas production kinetics, to determine if blending of ruminal solids (which increases blank gas production) improves the rate of gas production kinetics, and to determine if preparing inoculum with warm buffer affects early chemical change by fermenting.

METHODS AND MATERIALS

Transistorized pressure transducers were used with a computer data acquisition system to continuously monitor pressure accumulation in sealed serum bottles with a head space for gas of about 48 ml after media is included. Approximately.1100 gm of material was fermented. Bottles containing the sample, a small Teflon-coated stir bar, 8 ml of media were warmed in a water bath at 39 °C. Each was individually purged with CO2 before. 3 ml of reducing solution was included and bottles were lightly stoppered. At the time of inoculation, bottles were moved from the water bath, injected with about 2 ml of ruminal fluid (or 4.1 ml of blended ruminal inoculum), and the inoculum weight was recorded. Each bottle was stoppered, sealed with an aluminum crimp, immediately connected to a transducer using a 20-gauge needle. Gas pressure was recorded every 0.1 hr for at least 3 hr. Sample materials added glucose, purified starch and citrus pectin, an alfalfa hay standard, and 20 alfalfa genotypes.

Rumen contents were collected in the morning and afternoon on three days from two lactating cows that were fed a mixed ration containing alfalfa silage, corn silage, high moisture corn, soybean meal and minerals. Cows were fed ad libitum with feed offered once in the morning of each day. Within 10 min after collection, the top 3-5 cm of solids were discarded and the remaining contents were strained through 2 layers of cheesecloth. In experiment 1, 50 gm of strained solids and 100 ml of strained ruminal fluid from each cow were combined and blended for 60 sec. The ruminal fluid and blended material were each strained through four layers of cheesecloth to obtain strained ruminal fluid (SRF) and blended ruminal fluid (BRF) that was used as inocula. Both SRF and BRF were also centrifuged at 500g for 5 min to produce centrifuged SRF (CSRF) and BRF (CBRF). In experiment 2, 50 ml of strained ruminal solids from each cow were combined with 220 ml of previously warmed and reduced in vitro media and blended for 60 sec. Blended material and 100 ml of strained ruminal fluid from each cow were strained through four layers of cheesecloth to obtain blended ruminal media (BRM).

RESULTS AND DISCUSSION

The dramatic depression in gas production when ruminal fluids were centrifuged was also unexpected. The low centrifugal force that was used could not have segmented bacteria unless they were attached to particles. It is suspected that the reduced activity of centrifuged inocula may be due to temperature decreases during centrifugation. The heat of air of these inocula were about 33 °C at the time of inoculation. Temperatures of BRF were also less than 39°C at the time of inoculation due to cooling of the ruminal fluid during blending.

In experiment 2, ruminal solids were blended with warmed in vitro media that had been purged with CO2 and reduced. It was hypothesized that starting of solids with warmed media would detach additional ruminal micro-organisms and improve initial rates of gas production. BRM resulted in slightly higher blank gas production, but also significantly become greater gas production of samples. In both experiments, inocula collected in the afternoon had higher blank gas production, but the gas production of samples was higher in the PM in experiment 1 and lower in the PM in experiment 2. There were differences among days, but the day effect could be removed by using a standard hay or purified pectin as a covariate. In experiment 2 we observed an effect due to location in the incubator that is unexplained, but could be accounted for statistically.

SUMMARY AND CONCLUSION

It appears that both the source and method of preparing inoculum can affect in vitro gas production during early chemical change by fermenting. To maximize the rate of gas production and ensure that the method used in the in vitro system is not limiting digestion kinetics, ruminal solids should be blended with warmed and reduced in vitro media to extract additional micro-organisms. It is recommended that ruminal contents be collected from at least 2 and preferably 4 cows fed different diets and that strained ruminal fluid from these cows be mixed with warmed, reduced media that has been blended with ruminal solids and strained.

Brewer's yeast

Brewer's forth appearing in fermentation is made from a one-celled fungus called *Saccharomyces cerevisiae* and is used in making beer. It also can be grown specifically to make nutritional supplements. Brewer's forth appearing in fermentation is a rich source of minerals — particularly chromium, an essential trace mineral that helps the body maintain normal blood sugar levels; selenium; protein; and the B-complex vitamins. It tastes bitter and should not be confused with baker's yeast, nutritional yeast, or torula yeast; all those types of yeast are low in chromium. Brewer's yeast has been used for years as a nutritional supplement.

GENERAL

Brewer's forth appearing in fermentation is often used as a source of B-complex vitamins, chromium, and selenium. The B-complex vitamins in brewer's yeast include B1 (thiamine), B2 (riboflavin), B3 (niacin), B5 (pantothenic acid), B6 (pyridoxine), B9 (folic acid), and H or B7 (biotin). These vitamins help break down carbohydrates, fats, and proteins, which provide the body with energy. They also support the nervous system, help maintain the muscles used for digestion, and keep skin, hair, eyes, mouth, and liver healthy. Above all, brewer's yeast does not contain vitamin B12, an essential vitamin found in meat and dairy products; vegetarians sometimes take brewer's yeast mistakenly believing that it provides B12, which can be lacking in their diet.

Yeast Metabolism

Metabolism refers to the biochemical assimilation (anabolic pathways) and dissimilation (catabolic pathways) of nutrients by a cell. Like in other organisms, in forth appearing in fermentation these processes are mediated by enzymic reactions, and regulation of the underlying pathways have been studied in good detail in forth appearing in fermentation. Anabolic pathways add reductive processes leading to the production of new cellular material, while catabolic pathways are oxidative processes which remove electrons from substrates or intermediates that are used to generate energy. Preferably, these processes use NADP or NAD, respectively, as co-factors.

Although all yeasts are micro-organisms that derive their chemical energy, in the from of ATP, from the breakdown of organic compounds, there is metabolic diversity in how these organisms generate and consume energy from these substrates. Knowledge of the underlying regulatory mechanisms is not only valuable in the understanding of general principles of regulation but also of great importance in biotechnology, if new metabolic capabilities of individual yeasts have to be exploited. It is now well established that most yeasts employ sugars as their main carbon and hence energy source, but there are particular yeasts which can utilize non-conventional carbon sources. With regard to nitrogen metabolism, most yeasts are capable of assimilating simple nitrogenous sources to biosynthesise amino acids and proteins. Aspects of phosphorus and sulphur metabolism as well as aspects of metabolism of other inorganic compounds have been studied in some detail, predominantly in the forth appearing in fermentation, Saccharomyces cerevisiae.

SUGAR CATABOLISM IN YEAST

The major source for energy production in the forth appearing in fermentation, Saccharomyces cerevisiae, is glucose and glycolysis is the general pathway for conversion of glucose to pyruvate, whereby production of energy in form of ATP is coupled to the generation of intermediates and reducing power in form of NADH for biosynthetic pathways.

Two principal modes of the use of pyruvate in further energy production can be distinguished: respiration and chemical change by fermenting. In the presence of oxygen and absence of repression pyruvate enters the mitochondrial matrix where it is oxidatively decarboxylated to acetyl CoA by the pyruvate dehydrogenase multi enzyme complex. This reaction links glycolysis to the citric acid cycle, in which the acetyl CoA is completely oxidised to give two molecules of CO2 and reductive equal in form of NADH and FADH2. Above all, the citric acid cycle is an amphibolic pathway, since it combines both catabolic and anabolic functions. The latter results, for example, from the production of intermediates for the synthesis of amino acids and nucleotides. Replenishment of compounds necessary to drive the citric acid cycle, such as oxaloacetate and á-ketoglutarate, are (i) the fixation of CO2 to pyruvate by the actions of the enzymes pyruvate carboxylase (ATP-dependent) and phosphoenolpyruvate carboxykinase and (ii) the glyoxalate cycle which is important when yeasts are grown on two-carbon sources, such as acetate or ethanol. During alcoholic chemical change by fermenting of sugars, yeasts reoxidise NADH to NAD in a two-step reaction from pyruvate, which is first decarboxylated by pyruvate decarboxylase followed by the reduction of acetaldehyde, catalysed by alcohol dehydrogenase (ADH). Concomitantly, glycerol is generated from dihydroyacetone phosphate to ensure production of this important compound.

The mode of glucose oxidation is the hexose phosphate pathway also known as the pentose phosphate cycle, which provides the cell with pentose sugars and cytosolic NADPH, necessary for biosynthetic reactions, such as the production of fatty acids, amino acids and sugar alcohols. The first step in this pathway is the dehydrogenation of glucose-6-phosphate to 6- phosphogluconolactone and generation of one mole of NADPH (glucose-6-phosphate dehydrogenase). Subsequently, 6-phosphogluconate is decarboxlated by the action of phosphogluconate dehydrogenase to give ribulose-5-phosphate and a second mole of NADPH. Thus, besides generating NADPH, the other mayor function of this pathway is the production of ribose sugars which serve in the biosynthesis of nucleic acid precursors and nucleotide coenzymes.

NAD and FAD, which become reduced during the breakdown of sugars to NADH and FADH2, respectively, are reoxidized in the respiratory (electron transport) chain located in the inner mitochondrial membrane. The energy released during the transfer of electrons is coupled to the process of oxidative phosphorylation, which is effected by ATP synthase, an enzyme complex which is also located in the inner mitochondrial membrane and designed to synthesize ATP from ADP and inorganic phosphate.

ETHANOLASSAY

Method

Most of the chemical oxidation methods are based on the complete oxidation of ethanol by dichromate in the presence of sulfuric acid with the chemical change by fermenting of acetic acid. This reaction is popular because potassium dichromate is easily available in high purity and the solution is indefinitely stable in air. The theoretical reaction stoichiometry is shown below:

2Cr2O7— + 3C2H5OH + 16H+ ——> 4Cr+++ + 3CH3COOH + 11H2O

Dichromate (Cr2O7—, Cr(VI)) is yellowish in colour and the reduced chromic product (Cr+++, Cr(III)) is intensely green. Because the absorption spectra of dichromate and chromic ions overlap significantly, Beer's law is not obeyed. Instead, the spectra of the solution of interest must be analyzed at multiple wavelengths to calculate the particular concentrations of dichromate and chromic ions in a mixture subject to the material balance that the total number of chromium atoms must be conserved.

Other methods of determination primarily based on the above reaction are commonly used. In these methods, any one of the reactants or products participating in Reaction (1) can be analysed through another separate reaction. For e.g., the excess dichromate remaining in the solution can be further reduced by titration with other oxidizing reagents such as ferrous ammonium sulfate:

Cr2O7— + 6Fe++ + 14H+ ——> 2Cr+++ + 6Fe+++ + 7H2O

To enhance the visualisatio of titration endpoint, organic indicators such as sodium diphenylamine sulfonate and 1,10-o-phenanthroline are added. Another similar method is based on iodometric titration.

Proper concentration of sulfuric acid in the surronding solution will direct the oxidation of ethanol towards acetic acid instead of acetaldehyde. Or the excess dichromate can be analysed.

7

Inoculum Preparation and Strain Improvement

In order to provide a more seamless transition of basic research into the clinic there is a pressing need to increase the reliability and physiological relevance of early screening campaigns to determine true lead compounds. Informal Life Sciences' Primary and Secondary Screening congress will set the stage for one of the key screening events of the year.

Co-located for the second year running with our Compound Library Design, Synthesis and Acquisition conference, a joint plenary session will bring together chemists and screening scientists. This will provide the rare opportunity to examine the needs and capabilities of each to look and present examples where the strengthening of the relationship can have a significant collision on the timelines and success of hit and lead discovery and optimisation phases.

The primary screening programmers serves to remove worthless microorganisms on the basis of relatively simple fermentation criteria. The activity of antibiotics in vitro and possibly in via, against a small number of the most important fast organism.

For e. g. —micro-organisms capable of producing organic acids, amino, vitamins.

CROWDED PLATE TECHNIQUE

The crowded plate technique is the simplest screening technique employed in detecting ad isolating a series of dilution of the soil or other source material for the antibiotic producing microorganisms followed by spreading of dilution on the nutrient agar plated. The ability of a colony to exhibit antibiotic activity is indicated by the presence of ozone of growth of inhibition surrounding the colony.

The crowded plate technique has a limited application, since it merely provides information regarding the inhibitory activity of a colony against the unwanted microbes that may be present by change on the plate. Then the plates are subjected to further incubation to allow the growth of the test organism. The formation of inhibitory zones around certain colonies indicated their antibiotic activity.

Auxanography is a technique which is largely employed for detecting microorganisms able to produce growth factors. The two major factor of the technique are given.

Preparation of first plate

(i) A filter paper strip (15x12cm) is put across the bottom of a peal dish in such a way that the two ends pass over the edge of the dish.
(ii) The filter paper disc of Petri dish size is placed over paper strip on the bottom of the dish.
(iii) The nutrient agar is pound on the paper disk.
(iv) Microbial source material.
(v) Plating of aliquots properly diluted soil sample is done.

Preparation of second plate

(i) A minimal medium lacking the growth factor under consideration is seeded with the test organism.
(ii) The seeded medium is pound on the surface of fresh Petri disk.
(iii) The plate is allowed to set.

Enrichment Culture Technique

ENRICHMENT CULTURE

Enrichment culture is a very valuable technique developed by some giants in microbiology in the early-1900s. The basic principle involved is that of selection. Often, it is desired to isolate bacteria that are relatively scarce, or are in fact in very low numbers. This creates a problem. Normally, bacteria are isolated from one another using the streak plate technique, in which a sample of bacteria is spread out onto an agar plate in such a way that particular cells are separated from each other. Each cell multiplies to form a visible colony in which all the cells are the same strain. Thus you can use a loop to pick some cells from a colony to start a pure culture, a culture in which a single organism is growing. If the desired organism is a minority in the sample, say 0.1 per cent of the total, one would have to have at least a thousand isolated colonies on a plate to have a chance of seeing just one of the desired bacterium. This is physically impossible; there is not sufficient room on a Petri dish.

Enrichment culture solves this problem. The essence of this technique is to provide growth conditions that are very favourable for the organism of interest, and as unfavourable as possible for competing organisms. For e.g., if you wish to isolate a bacterium that is a thermophile incubate the sample at that high temperature. Organisms that cannot tolerate that temperature will dies or simply fail to grow, while thermophiles will grow and become greater in number, over time becoming a large and larger proportion of the total bacterial population in the sample. This is an example of enrichment by modifying the physical conditions.

Enrichment can also be carried out by modifying the nutrient content of the culture medium. Here are two examples. There are a variety of bacteria present in soil that are capable of converting the otherwise inert form of nitrogen in the atmosphere, N2, into ammonia. Such reactions provide the major input of N into the biosphere. Organisms capable of this transformation of N can be isolated by incubating a soil sample in a culture medium which has all the ingredients necessary for growing except nitrogen. Those bacteria that can "fix" nitrogen and create their own nitrogenous nutrients for growth will have a selective advantage and increase in number compared to those who can't. Nitrogen is likely to present in small amounts as a contaminant in the soil or in the other culture ingredients, so it would be incorrect to say that only nitrogen fixing bacteria will grow. They will, above all have a great advantage and will eventually outgrow the others given enough time.

Another example of manipulating the culture conditions would be an attempt to isolate bacteria that can break down a xenobiotic such as a pesticide. Pesticides are organic compounds. The compounds are C and H. Bacteria generally break down organic compounds as a source of building materials and energy. In a growth medium in which the pesticide in question is the sole carbon source for growth, only those bacteria that can use the compound will grow appreciably, while others will not. Eventually, the culture will contain a sufficiently high proportion of pesticide-degraders that it would be easy to isolate them using the streak plate technique.

Microbiologists sometimes argue about what should correctly be called enrichment culture. A commonly used method in molecular genetics is to select for bacteria which carry a plasmid with an antibiotic resistant gene by growing that bacterium in a medium containing the antibiotic. This type of selection really isn't what we mean by enrichment culture.

You may have an opportunity to carry out this technique. The slants of igepon-degrading bacteria that we are studying were isolated in this way.

THEORY

pH indicators themselves are frequently weak acids or bases. The general reaction scheme of a pH indicator can be formulated as follows:

HInd + H2O Ì! H3O+ + Ind-

Here HInd stands for the acid form and Ind- for the conjugate base of the indicator. It is the ratio of these that determines the colour of the solution and that connects the colour to the pH value. For pH indicators that are weak protolytes, we can write the Henderson-Hasselbalch equation for them:

The equation, derived from the acidity constant, states that when pH equals the pKa value of the indicator, both species are present in 1:1 ratio. If pH is above the pKa value, the concentration of the conjugate base is greater than the concentration of the acid, and the colour associated with the conjugate base dominates. If pH is below the pKa value, the converse is true.

Usually, the colour change is not instantaneous at the pKa value, but there is a pH range where a mixture of colours is present. This pH range varies between indicators, but as a rule of thumb, it falls between the pKa value plus or minus one. This assumes that solutions retain their colour as long as at least 10 per cent of the other species persists. For e.g., if the concentration of the conjugate base is ten times greater than the concentration of the acid, their ratio is 10:1, and consequently the pH is pKa + 1. Conversely, if there is a tenfold excess of the acid with respect to the base, the ratio is 1:10 and the pH is pKa – 1.

For optimal accuracy, the colour difference between the two species should be as clear as possible, and the narrower the pH range of the colour change the better. In some indicators, such as phenolphthalein, one of the species is colourless, whereas in other indicators, such as methyl red, both species confer a colour. While pH indicators work efficiently at their designated pH range, they are usually destroyed at the extreme ends of the pH scale due to undesired side-reactions.

APPLICATION

pH indicators are frequently employed in titrations in analytic chemistry and biology experiments to determine the extent of a chemical reaction. Because of the subjective determination of colour, pH indicators are susceptible to imprecise readings. For applications requiring precise measurement of pH, a pH metre is frequently used.

Tabulated below are several common laboratory pH indicators. Indicators usually exhibit intermediate colours at pH values inside the listed transition range. For e.g., phenol red exhibits an orange colour between pH 6.8 and pH 8.4. The transition range may shift slightly depending on the concentration of the indicator in the solution and on the temperature at which it is used.

COMMERCIAL PREPARATIONS

Universal indicator and Hydrion papers are blends of different indicators that exhibits several smooth colour changes over a very wide range of pH values.

NATURALLY OCCURRING PH INDICATORS

Many plants or plant parts contain chemicals from the naturally-coloured anthocyanin family of compounds. They are red in acidic solutions and blue in basic. Extracting anthocyanins from red cabbage leaves or the skin of a lemon to form a crude acid-base indicator is a popular introductory chemistry demonstration.

Anthocyanins can be extracted from a multitude of coloured plants or plant parts, including from leaves (red cabbage); flowers (geranium, poppy, or rose petals); berries (blueberries, blackcurrant); and stems (rhubarb). An exhaustive list would be beyond the scope of this article.

Mineral oil

Mineral oil or liquid petroleum is a by-product in the distillation of petroleum to produce gasoline and other petroleum based products from crude oil. It is a transparent, colourless oil composed mainly of alkanes (15 to 40 carbons) and cyclic paraffins, related to petroleum jelly. It has a density of around 0.8 g/cm3. Mineral oil is a substance of relatively low value, and it is produced in very large quantities. Mineral oil is available in light and heavy grades, and can often be found in drug stores.

In the late-1800s, the term "mineral oil" or "rock oil" was first used to describe the petroleum hydrocarbons and associated products that were produced from wells that tapped underground reservoirs. The term differentiated petroleum hydrocarbons produced from underground sources from other common oil sources at the time, such as palm oil or whale oil. In today's petroleum exploration and production (E andP) business, the phrase "mineral oil" is most often used in legal documents to define and encompass all of the liquid hydrocarbon and gaseous products produced from wells drilled into underground petroleum-bearing reservoirs.

There are three basic classes of refined mineral oils:

(1) Paraffinic oils, based on n-alkanes,
(2) Naphthenic oils, based on cycloalkanes, and
(3) Aromatic oils, based on aromatic hydrocarbons (not to be confused with essential oils).

Applications

Mineral oil with included fragrance is marketed as baby oil in the US, UK and Canada. While baby oil is primarily marketed as a generic skin ointment, other applications exist in common use. It is often used on infant "diaper rashes" to ease the inflammation. It may alleviate mild eczema, particularly when the use of corticosteroid creams is not desirable. Mineral or baby oil can also be employed in small quantities (2–3 drops) perday to clean inside ears. Over a couple of weeks, the mineral oil softens dried or hardened earwax so that a gentle flush of water can remove the debris. In the case of a damaged or perforated eardrum, above all mineral oil should not be used, as oil in the middle ear can lead to ear infections. Mineral oil is used as a suspending and levigating agent in sulphur-based ointments.

Medicine – Internal Uses

Mineral oil is taken orally as a lubricative laxative, and is often prescribed to ease the pain of bowel movements for those who suffer from hemorrhoids and constipation. In Europe, the use of mineral oil as laxative is considered obsolete mainly due to its potentially harmful effects on the lungs if accidentally aspirated. Furthermore, the oil may be absorbed to a small percentage into internal tissue and cause adverse reactions to the body. In higher therapeutic dosages loss of bowel control and/or dripping from the rectum has been reported causing temporary stool incontinence. Mineral oil temporarily coats the intestines and prevents the uptake of certain essential vitamins and nutrients. Mineral oil is never to be used in conjunction with other stool softeners such as sodium docusate as this become greater the likelihood of harmful absorption of mineral oil into the body.

Veterinary Medicine

Certain mineral oils are used in livestock vaccines, as an adjuvant to stimulate a cell-mediated immune response to the vaccinating agent. In the poultry industry, plain mineral oil can also be swabbed onto the feet of chickens infected with scaly mites on the shank, toes, and webs. Mineral oil suffocates these tiny parasites. In beekeeping, meal grade mineral oil saturated paper napkins placed in hives are used as a manner of appling remedies for tracheal and other mites.

Cosmetics

Mineral oil is a common ingredient in baby lotions, cold creams, ointments and cosmetics. It is a lightweight inexpensive oil that is odorless and tasteless. It can be used on eyelashes to prevent brittleness and breaking and, in cold cream, is also used to remove creme makeup and temporary tattoos. One of the common concerns regarding the use of mineral oil is its presence on several lists of comedogenic substances. These lists of comedogenic substances were developed many years ago and are frequently quoted in the dermatological literature.

Mechanical, Electrical and Industrial

Mineral oil is used in a variety of industrial/mechanical capacities as a non-conductive coolant or thermal fluid in electric components as it does not conduct electricity, while simultaneously functioning to displace air and water. Some examples are in transformers where it is known as transformer oil, and in high voltage switchgear where mineral oil as an insulator and as a coolant to disperse switching arcs. The dielectric constant of mineral oil ranges from 2.3 at 50 degrees Celsius to 2.3 at 200 degrees Celsius.

Electric space heaters sometimes use it as a heat transfer oil. Because it is non-compressible, mineral oil is used as a hydraulic fluid in hydraulic machinery and vehicles. Mineral Oil is also used as a lubricant. Light mineral oil is also used in textile industries and used as a jute batching oil. An often cited limitation of mineral oil is that it is poorly biodegradable; in some applications, vegetable oils such as cottonseed oil or rapeseed oil may be used instead.

Preservative

Since it does not absorb atmospheric moisture, mineral oil is useful as a protective coating or bath for water-sensitive materials. Alkali metals like lithium are often submerged in mineral oil for storage or transportation.

Mineral oil is also often used as a coating on metal tools and weapons, knives in particular, as a way to inhibit oxidation. The Japanese swords Nihonto, for example, are traditionally coated in Choji oil which consists of 99 per cent mineral oil and 1 per cent oil of cloves. The use of oil of cloves is sometimes explained as a means of differentiating sword oil from cooking oil to prevent accidental ingestion, but may also be purely aesthetic.

Mineral oil can be used as a leather conditioner as well, though most shoe polishes use naphtha, lanolin, turpentine and Carnauba wax instead.

It can also be used as a wood preservative. A light coating of mineral oil, rubbed into well-sanded wood, provides an easy-to-apply and relatively durable finish, without the odor or drying time of varnish or urethane.

Meal Preparation

Mineral oil's ability to prevent water absorption, combined with its lack of flavour and odor, make it a popular preservative for wooden cutting boards, salad bowls and utensils. Rubbing a small amount of mineral oil into a wooden kitchen item periodically will prevent absorption of meal odors and ease cleaning, as well as maintain the integrity of the wood, which is otherwise subjected to repeated wetting and drying in the course of use. The oil fills small surface cracks that may otherwise harbour bacteria.

It is occasionally used in the meal industry, particularly for candy. In this application, it is typically used for the glossy effect it produces, and to prevent the candy pieces from adhering to each other. It has been discouraged for use in children's meal, though it is still found in many candies, including the popular movie theater treat Swedish Fish.

It can be used as a release agent for baking pans and trays, but meal oils like vegetable oil are a more popular choice.

Miscellaneous

Mineral oil's ubiquity has led to its use in some niche applications as well.

(1) Mineral oil can be used as a dust suppressant.
(2) In microbiology, mineral oil may be added atop agar stab growth media to create an anaerobic environment.
(3) It is the basis for most automotive engine oils.
(4) Used on larger diamond blades of lapidary slab saws as a coolant. Trim saws of 10" diamter or less can utilize water, often with an additive, but larger saws use mineral oil as it disapates the heat developed from cutting rock with diamond blades.

(5) It is used to make lava lamps.
(6) Mineral oil is used to darken soapstone countertops for aesthetic purposes.
(7) It is commonly used to create a "wear" effect on new clay poker chips, which can otherwise only be accomplished through prolonged use. The chips are either placed in mineral oil or the oil is applied to each chip individually, then rubbed clean. This removes any chalky residue leftover from manufacture, and also improves the look and "feel" of the chips.
(8) It has a high refractive index, so it is sometimes used in oil immersion microscopes.
(9) It is the principal fuel in some types of gel-type scented candles.
(10) It is an effective pesticide, particularly for edible plants. It is effective against a wide range of insects and all stages of insect development.
(11) Mineral oil has been used to immerse computers in order to absorb heat and cool the system in some custom-built projects.
(12) It is sometimes used as a personal lubricant and as an alternative to plant or herbal oils for massage.
(13) It can be used in some model trains as a substitute for the "smoke fluid" or "smoke oil" that simulates steam coming from a steam engine.
(14) It can be used in basement floor drain traps to float on top of the water slowing its evaporation thereby keeping sewer gas from entering the house for a longer period of time.

Other names

The broad range of applications for mineral oil has resulted in an equally expansive list of application-specific names and trade brands. Other names for mineral oil include:

- Heat-treating oil;
- Hydraulic oil;
- Lignite oil;
- Liquid paraffin;
- Mineral Seal Oil;
- Paraffin oil;
- Petroleum, liquid;
- White oil;
- Adepsine oil;
- Albolene;
- Cable oil;
- Baby Oil;
- Drakeol; and
- Electrical Insulating Oil.

8

Fermenter Design

The design of equipment has evolved in a largely empirical manner. The earliest s required only rudimentary standards of hygiene, either because of the nature of the substrate, or because the vigour of the desired organism exceeded that of potential competitors, or because the products of were inhibitory, or because the expected shelf- life of the product was short. This applied to the fermentation processes involved in the make the product by machine of wine, beer, cheese, youghourt, vinegar, sauer- kraut and so on. Even with these products the transition from domestic art to commercial practice required improved standards to increase shelf-life and to maintain acceptably-consistent standards of product quality, hut equipment and process control remained essentially simple. Some sophistication occurred with the development of pure-culture techniques in beer-making, but the first really fastidious, the make the product by machine of acetone and butanol, was initiated on a large scale less than sixty years ago. In this system the maintenance of a strictly anaerobic surrounding object was essential. This provided protection against a wide range of contaminants requiring atmospheric oxygen, and the principal hazard was contamination by bacteriophage.

The next mayor development was the adaptation of submerged-culture techniques for the production of penicillin, rapidly followed by processes for other antibiotics, vitamins, and amino acids, and the conversion of steroids. These processes have been developed over the past thirty years or so, and have attracted the combined attention of microbiologists. hiochemists and engineers to resolve a variety of problems in order to improve both submit and process efficiency. More recently, a best deal of attention has been devoted to problems associated with the replacement of batch processes by continuous processes, and the utilisation of gaseous and liquid hydrocarbons as substrates, instead of the traditional carbohydrates. The latter processes, in particular, have stimulated investigations into the use of fermenter configurations other than the agitated cylindrical vessel which has been almost universally adopted for aerobic microbial processes and for some anaerobic processes. Among the new configurations two in particular—the air-lift and the tower—have been exploited commercially and have attracted the attention of research workers.

There are two types of fermenters with reference:

1. Pilot fermenters.
2. Large fermenter.

The pilot fermenters are used in large state studies of fermentation. They have size of twenty five to hundred gallons and con go upto 2000 gallons of total volume.

Large fermenter is used in industrial production of fermentation product of microbial cells. They have size of 5000 to 10,000 gallons.

1. Aubmerged culture design
   a. STR system
   b. Tower fermenter
   c. Activated sludge process
   d. Solid substrate fermenter

Advantage

(i) Superior productivity;
(ii) Simple technique;
(iii) Law capital investment;
(iv) Reduced energy requirement;
(v) Low waste water out put; and
(vi) No foam problem.

limitation

(i) Heat build-up; and
(ii) Bacterial controlling substrate moisture level.

FERMENTER DESIGN AND THE REQUIREMENTS OF THE MICROBIAL SYSTEM

The function of the fermenter is to provide an surrounding object suitable for the controlled growth of a pure culture or of a defined mixture of organisms. The materials of construction must be such that they will not adversely affect, nor be adversely affected by, the desired microbial activity, either by interaction with the medium or by harbouring unwanted organisms. They must be resistant to corrosion by the nutrient medium and products, and to the effects of sterilisation temperatures. The actual con—struction of the equipment from suitable materials must also take account of these factors and of the stresses imposed by pressurisation and the weight of the vessel contents. There must be provision for the regulation of heat of air and of the supply of air, for charging and discharging the vessel contents. for inoculation, and generally also, for sampling and for the control of pH and foaming. Frequently, even in batchwise systems, it will be necessary to provide for controlled addition of nutrients or other materials during the course of the chemical change by fermenting

In continuous-culture systems additional facilities must be provided to control culture volume and medium flow-rate. Depending on the method of control, some system will be required to measure and control the cell concentration or the concentration of some rate-limiting substrate. These controls relate to the macroenvironment and only measure or control the conditions of the micro-organisms in an indirect, empirical sense. Some control systems, such as those based on the rate of oxygen uptake, evolution of Co2 or concentration of NADH, relate more directly to cell activity, but none gives a direct measure of the condition in the microenvironment, that is the surrounding object in the immediate vicinity of the cell.

It is an assumption implicit in all control systems that biochemically/—similar macroenvironments produce biochemically-similar microenvironments. Anomalies and difficulties which arise in scaling-up or in conducting a given chemical change by fermenting in different types of equipment indicate that this assumption may be false.

One possible source of anomalies is differences in the extent of cell aggregation, which, in biochemically-similar surrounding object, may be strongly affected by hydrodynamic factors. Another likely source of error lies in systems for assessing conditions in the macroenvironment.

These will generally indicate a time-average value at a particular sampling or sensing point and may conceal considerable variations from one element of culture to another. Thus attempts to control conditions in the culture as a whole may have somewhat attenuated effects on the microenvironment. At the same time, the activity of the micro-organism will itself have effects on the microenvironment, both directly and through effecting changes in the macroenvironment. The direct effects on the microenvironment will result from: (1) consumption of substrates, (2) output of products, (3) production of heat, and (4) aggregation. Factors (1) and (2) will result in deficiency of substrate and accumulation of product, with the possibility of resultant inhibition, unless there is adequate interchange between the region immediately adjacent to the cell and the bulk of the fluid. A similar interchange is also necessary to prevent significant and possibly damaging become greater in temperature. Aggregation of cells will adversely affect both heat and mass transfer, partly by reducing the area available for transfer, partly by becoming greater the length of the transfer paths and partly by reducing the transfer coefficients.

The effects of the organism on the macroenvironment, which may be reflected in concomitant changes in the microenvironment, arise mainly from (a) changes in the rheological character of the culture and (b) changes in interfacial tension. Changes in interfacial tension arise as a result of changes in the composition of the medium owing to the metabolic activities of the organism, and may affect both effective mass-transfer coefficients and the state of dispersion of the microorganisms and gas bubbles.

Rheological changes may result from the breakdown of macromolecular substrates, the formation of extracellular macromolecular products or simply from interference between cell aggregates. This latter effect is particularly pronounced for filamentous organisms and frequently results in non-Newtonian behaviour. These changes may have marked on rates of mass transfer and on the quality of mixing.

A mayor feature of the stirred vessel over other designs and one which may largely account for its popularity, is the degree of operational flexibility which it provides even when installed. This arises largely because mixing and mass transfer are influenced both by the action of the impeller(s) and by the rate of aeration, which can, within fairly wide limits, be varied independently, albeit at the expense of changing the impeller speed or geometry. By contrast, in the air-lift mixing and mass transfer are both dependent in a given piece of equipment, on the rate of aeration, and cannot readily be varied independently. In the tower fermenter, unless provision is made for recycling, mixing and mass transfer are strongly affected by the dilution rate, which must be determined on biological grounds, whereas the dilution rate can be fixed independently of considerations of aeration and mass transfer in stirred vessels and air-lift fermenters.

CHEMICAL ENGINEERING ASPECTS OF FERMENTER DESIGN

The objective of fermenter design and operation is to ensure that the desired activity of the microorganisms concerned shall not be restricted by the characteristics of the equipment. It is, so, useful to consider the problem in terms of the physical processes which might limit microbial activity.

Mass Transfer

The transfer of mass within the system is fundamental to the whole operation. There are two aspects to this process—the more-or-less uniform distribution of substrate and product molecules in the bulk of the fluid, and transfer between the bulk of the fluid and the microbial cells. The first aspect is largely governed by the forced convection and turbulence produced by agitation and the flow of gas, whereas transfer between the bulk and cells is determined by diffusional forces. It is not necessary that there should be a uniform concentration of any given component in the bulk of the fluid: only that it should not at any point fall below or rise above the level at which the desired level of microbial activity will be reduced. With materials whose solubility is many times greater than the rate-limiting concentration this generally presents no problem, although this must not be assumed to be the case in continuous flow chemostats where control is exercised by deliberately maintaining the concentration of a selected substrate at a rate-limiting level. For materials whose solubility is relatively close to the limiting concentration uniformity of distribution is more momentous. This could apply, for instance, to mineral salts of low solubility and, perhaps more importantly, to substrates such as steroids and hydrocarbons. Most importantly it applies to oxygen. When supplied in a stream of air, as is usually the case, the equilibrium concentration in fermentation media is only of the order of five to ten times the limiting concentration, even in the absence of microbial activity. In an actively respiring culture this margin is further reduced. Moreover, the amount in 'suspension' is limited by the gas hold-up in the system, a limitation which does not apply to the same extent to liquid substrates and hardly at all to solid substrates.

For particles of a given size and density in a specified fluid the available evidence suggests that the specific mass-transfer coefficient is unaffected by the intensity of agitation provided that all particles are freely suspended. Calderbank and Jones' have shown that, for systems in which the density difference between the continuous and dispersed phases is small, the specific mass-transfer coefficient is inversely proportional to the diametre of the suspended particles. This will apply to a suspension of microorganisms in an aqueous medium, and suggests that there will be a beneficial effect from levels of agitation which prevent aggregation of the microorganisms. There will be a further feature in that increasing the degree of dispersion will become greater the interfacial area available for transfer between a given mass of organisms and the suspending medium.

If the density difference between the dispersed and continuous phases is large, the specific mass-transfer coefficient is generally independent of particle diametre. In the case of gas bubbles, however, two regimes can be distinguished. Small bubbles behave as rigid spheres, whereas large bubbles can be distorted by hydrodynamic forces. In a given gas—liquid system the specific mass-transfer coefficient for large bubbles is greater than that for small bubbles, but in each region the coefficient is unaffected by the actual size and intensity of agitation. This suggests that any effect of agitation on the volumetric mass-transfer coefficient must arise from increases in surface area as a result either of reducing bubble diametre within the regime concerned or of increasing gas hold-up by recirculation and back-mixing. It is generally assumed that stirred, spragged chemical change by fermenting systems operate in the small-bubble regime but recent results obtained in our laboratories suggest that some systems may operate in the transition region. If this is the case, both absolute and volumetric mass-transfer rates will be very unpredictable and this may lead to difficulties both in scaling-up and in comparing different types of equipment.

Heat Transfer

Heat is generated in submerged microbial systems and partly by the metabolic activity of the organisms and partly as a result of the mechanical work performed by the agitator and by the gas-bubbles as they expand in passing from the sparger to the head-space above the culture. In mechanically—agitated systems the work performed by gas expansion is usually a small proportion of the total.

Some heat is lost as a result of become greater humidification of the gas stream in its passage through the culture and some by radiation from the outer surfaces of the fermenter. In small vessels these losses may be so large as to make it necessary to supply additional heat to maintain the desired heat of air. As the scale of operation become greater, the situation with regard to heat dissipation changes. Frequently, the relative volumetric gas-flow rate will be reduced, thereby reducing relative loss of heat by evaporation. More importantly, for geometrically similar vessels the volume increases as the third power of the linear dimensions, whereas the surface area become greater only as the second power. As a result, a scale is reached at which the loss of heat by evaporation and radiation is insufficient to prevent the temperature rising above the desired level. It is then necessary to provide additional cooling, either by passing a coolant over or around the wall of the vessel or by fitting cooling coils inside the vessel. This then serves to remove heat by conduction.

The transfer of heat between the micro-organisms and the surface of the vessel is somewhat analogous to that of mass transfer, involving interchange between the cells and bulk fluid and between the bulk fluid and the cooling surfaces. As in the case of mass transfer, the temperature in the bulk fluid is determined by a dynamic balance between the processes of heat evolution by the cells and agitator and of heat removal through the vessel walls. The direct determination of cell-to-fluid heat-transfer coefficients is much more difficult than the determination of mass-transfer coefficients, but it can be anticipated that the accumulation of heat within cell aggregates at a given level of metabolic activity will become greater as the size of the cell aggregates become greater. Thus, for a given temperature in the bulk fluid the average temperature of a cell will be expected to increase with increasing aggregation. In one momentous respect the process of heat transfer differs from that of mass transfer. Whereas mass transfer takes place throughout the whole of the dispersion, the interchange of heat with the surroundings takes place at well-defined and fixed boundaries. The effects of hydrodynamic factors are more definable in this situation. In individual changes in the velocity of the fluid over the heat-transfer surfaces may have marked effects on the rate of heat transfer, and such changes may occur as a result of changes in the rheological character of the culture. Another effect of biological activity is the fouling of heat-transfer surfaces, either by attack by materials in solution or by deposition of cells. Such fouling reduces the overall heat-transfer coefficient. Apart from the choice of constructional materials, only the size and positioning of baffles will substantially affect the situation at the vessel walls. The design, location and supporting of coils, above all, should be such as to minimize the risk of deposition by ensuring free flow of the suspension over the surface of the coils.

Mixing Effects

Mixing serves to minimize local variations in concentration and temperature. It arises from the random redistribution of elements of the culture suspension in the impeller zone, and from the random interaction, with the bulk of the suspension, of streams leaving the impeller. In a given system the rate of mixing is a direct, though not necessarily linear, function of the mass flow-rate from the impeller. This may be expressed as a mean circulation rate, that is the number of times that a volume of suspension equal to the total volume passes through the impeller zone in unit time. Alternatively, one may consider the mean circulation time, that is the time that is required for a volume equal to the total volume of culture to pass through the impeller zone. This will represent also the average time which will elapse between two successive passages through the impeller zone of a given element of suspension. Because of the random nature of the processes, the actual circulation times for a series of elements will be distributed about this mean.

For batch systems in which all particles are freely suspended, the distribution of circulation times is of little importance, since there will be adequate mass transfer throughout the suspension. The situation may be quite different, physical interference among cell aggregates frequently results in pseudoplastic behaviour. In the impeller region the particular aggregates are freely suspended in the nutrient medium, whilst retaining their integrity. As they pass into the bulk of the culture, they assemble into clusters as a result of experiencing less intense hydrodynamic forces. Mass transfer between these clusters and the bulk of the fluid may be much less than that characteristic of the individual aggregates. Consequently, nutrients available in limited amounts may become exhausted, or there may be excessive local accumulation of metabolic products, if such an assembly experiences a longer-than-average circulation time. This may result in reversible or irreversible loss of biological activity. Such effects may be particularly momentous in the case of O2 supply, since subjection of actively respiring organisms to conditions of oxygen deficiency may result in decrease of respiratory activity, or even a transition to anaerobic metabolism for facultative organisms.

If growth of the organism results in a marked become greater in viscosity and, more individually, if the system becomes pseudoplastic, these effects are accentuated by two factors— the establishment close to the impeller of a zone of low effective viscosity, through which much of the air will be channelled, and the establishment near the walls of the vessel of stagnant zones. Such zones may become permanently depleted of oxygen, resulting, at best, in reduced productivity, or more seriously, in lysis, anaerobiosis or other undesirable metabolic activities.

Each impeller produces two mixing zones, one above and one below the impeller level. Each of these zones can be considered to be well-mixed, but the extent to which the system as a whole can be considered to be well-mixed depends on the rate of interchange of material between the two zones. This effect and the formation of stagnant zones, mentioned earlier, have momentous implications for the location of sensors and sampling points, and arrangements for the addition and withdrawal of material during the course of a fermentation. These considerations are of particular importance in relation to continuous culture, especially in systems operating as chemostats, in which deviations from perfect mixing may lead to unpredictable performance and difficulty of control.

CONCLUSIONS

Biological activity depends basically on the microenvironment surrounding each cell. In practice, control of the microenvironment must be exercised rather indirectly through control of the overall conditions in the apparatus in which the process is conducted. It is clearly necessary to ensure that the overall supply of nutrients is adequate and that there should be provision for the removal of excess heat and of volatile products of metabolism, but this is insufficient to ensure optimal activity unless local variations in conditions can be kept within acceptable limits. Mechanical agitation is frequently employed to meet this latter need, but its effects are imperfectly understood and inadequately characterised. More needs to be known, for example, about the effects of mechanical forces on micro-organisms, the extent to which aggregation is determined by biological and mechanical factors and about the distribution of circulation times on various scales of operation, especially in non-Newtonian systems. The time, more information is required about the response of biological systems to transient variations in conditions which may be experienced while a particular element of culture suspension is circulated around the vessel under the action of the impeller. It may then be possible to assess quantitatively the extent, if any, to which desired biological activity is limited by variations in the physical conditions in a individual situation, and the benefits which might result from modification of those conditions.

This may be regarded as a distant and ideal objective, but benefits may accrue along the way. It is now possible to determine continuously many parametres in any biological system, and, with the aid of computers, to maintain the measured values within closer limits than has been possible by earlier methods. It does not follow that this has led to a correspondingly closer control of the microenvironment, nor that all the data which is processed is necessary to high-quality control. Fuller knowledge of physical conditions in microbial cultures and of their effects on biological activity will help to resolve these issues and to improve both equipment and process design. There seems no reason to believe that the spragged, agitated, baffled vessel will be displaced from its position of pre-eminence, but better characterisation may lead to better exploitation, and to the identification on rational grounds of situations in which radically different types of contacting equipment may be justified.

INTRODUCTION

The everybody conjures up the image of submerged fermentation (SF) as carried out in the big stainless steel fermentors or the fermentation vats. Not many are aware of the other type of fermentation called as solid substrate fermentations (SSF).

SSF is used in traditional food fermentations and it has always stayed in the back burners of the fermentation popularity. The 'unpopularity of SSF does not mean SSF is inferior. In certain ways SSF has more advantages to SF depending on the situation.

It is only in the last few decades that the West is interested in exploiting SSF in the production of microbial enzymes and other microbial products. It is interesting to point here that in terms of diversity of fermentation products the SSF is more restricted compared to SF

The unpopularity of SSF compared to SF is more attributed to the poor understanding and control of the SSF fermentation compared to SF which is well established in the West. The popularity of SF in the west is more because of the rich diversity of fermentation products that can be obtained through SF compared to SSF.

SF is more established in the west is probably initiated attributed to their fascination of alcoholic beverages. This is followed by the interest in the West to scale up or carry out industrial production through which they have through science and technology to understand and control the process. They would have recognized early that in terms of fermentation products SSF is more efficient compared to the passive SF

The jump start in SF is perhaps single handed contributed by the antibiotics fermentation industries during the World wars and see the legitimacy of Industrial Microbiology.

The poor understanding and limitations of SF is probably the main reason why SF always remain in the back burners of fermentation

It is comparatively, easier to carry out scale up of SF for industrial fermentations compared to SSF.

In a way it is wrong for us to regard SSF and SF as two different types of fermentations. Both types of fermentations are carried out by the micro-organisms.

The main differences between SSF and SF are that:

1 SSF uses little water.

2 The substrates for the microorganisms are significantly in solid forms and not in solution.

Little Water in SSF

When we say SSF uses little water in fermentation it could mean literally in terms of very low volume of water used in the fermentation or very little free or available water occurred in the fermentation.

This is a significant difference compared to SF where the aqueous phase is the dominant component in the fermentation process.

In situation where little water are used in SSF, water is still required fc metabolism of the fermentative microorganisms but it occur surrounding the substrate particles as a thin film of water. A good example is in tapai fermentation.

In Sauerkraut fermentation even though large volumes of water are used for the fermentation, these water are not free or easily available to the microorganisms due to competition with the salt ions.

Where very low volume of water are used in SSF, we could regard the SSF as dry fermentation as exhibited in tempe SSF where fungal mycelial growth covered the surfaces of the wet soya beans.

In such dry SSF, there are porous spaces around the sold substrate to allow for easy mass transfer of heat and oxygen in the substrate matrix.

Impact of Amount of Water in SSF

As discussed above water is the critical issue in SSF and SSF operate in very little water environment. Water is important in SSF as:

(1) It is the medium where nutrients required by the micro-organisms are dissolved and transported to the microorganisms.

(2) It is required by the microorganisms to carry out the various metabolic and biochemical reactions to grow

(3) The right amount of water is critical in any SSF. One good example is in the SSF of tempe. Higher amount of water could result in a more vigourous growth of the mycelia resulting in the clogging of the pores between the solid substrate and hampering the mass transfers of oxygen as well increase in the metabolic heat. This would result in a very hard and compact tempe which is not popular compared to the soft fluffy tempe.

(4) With the decrease in substrate porosity and the resulting decrease in oxygen may even increase the risks of bacterial contamination. Low amount of water on the other hand may result in poor accessibility of nutrients which will result in poor growth of the mycelia.

(5) Low amount of water in SSF means that the fermentation products in solution are not significantly diluted and will occur in high concentration.

Substrates in Solid Presentations

(1) In SF, the nutrients are usually dissolved, well mixed and dispersed in the fermentation broth. Good mixing will ensure good mass transfer of nutrients, oxygen and heat throughout the fermentor.

(2) In SSF, the conditions are significantly different as the substrates occur dominantly in particle or solid forms and are not easily available to the micro-organisms. These solid substrates are often static or fixed and not easily mixed or dissolved. SSF micro-organisms have to actively colonised the surface of the substrates, dissolved the solid substrates powered by extracellular enzymes to release the nutrients needed for the growth.

(3) The advantages of this dry SSF are that there are large surface areas available and easy aeration. Its setback is however low heat transfer capacity. However, often at times it is important in finding the correct size of solid particles in SSF. Too big a paricle might yield to surface area t volume ratio thus limiting the efficiency for microbial action. Yet at the same time using large particles will allow large void space to facilitate oxygen, water and heat mass transfer needed in aerobic SSF.

(4) In small particles SSF even though there is a higher surface area to volume ratio to optimise microbial action, it often however hinders the efficiency of the various mass transfer processes needed to support SSF

Aeration in SSF

(1) We are discussing aeration in the context of aerobic SSF. Aeration serves the following main functions in SSF.
   (a) to maintain aerobic conditions,
   (b) to desorb carbon dioxide,
   (c) to regulate the substrate temperature, and
   (d) to regulate the moisture level.

(2) In SSF that involves the formation of a thin film of water around the substrate the efficiency of oxygen mass transfer is very high in the case of dry SSF such as tempe. However, in tapai SSF due to the high sugar concentration resulting in the thick viscous film of liquid around the SSF, there is a very poor mass transfer of oxygen thus allowing alcohol fermentation to proceed

Heat Dissipations is Crucial Issue in SSF

(1) One of the main characteristics in dry SSF such as in tempe and composting is the high generation of heat generated by the aerobic metabolism. To make the conditions worst in SSF the solid materials used in SSF are often have low thermal conductivities. This would easily result in built up of heat.

(2) The effective dissipation of heat is often related to aeration of the SSF system. Aeration not only brings in oxygen but also help in the removal of heat from the SSF.

(3) The amount of heat produced and managed in SSF is very crucial to the fermentation process. High uncontrollable heat is not good in SSF as it affects the composition of micro-organisms, its physiology of growth and even product formation.

Micro-Organisms in SF and SSF

(1) In both type of fermentations, the micro-organisms are the agents of change. Differences might differ in the type of micro-organisms involved and the behaviour and physiological requirements of these micro-organisms. The function of the fermentors is to support the growth of the micro-organisms as such the type and

configuration of SF or SSF will be dictated by the nature of the micro-organisms involved

(2) In SSF the nature of microorganisms generally differ from SF as it involves:

a. Mixed cultures or diversity of micro-organisms.

b. That it does not rely on single or pure culture fermentations.

These two characteristics are consequences of using complex and solid substrate which can only be effectively utilised by the combined action of mixed microorganisms.

(4) Since it depend on mixed culture of micro-organisms which are in most cases natural inocula, aseptic is not really required in the SSF. Exclusivity of microorganisms in SSF in most cases are imposed by the phenomena of natural protection whereby not many organisms can grow in the environment such as the very low availability of free water or very high salt or sugar concentration

This does not mean to say SSF cannot go bad or contaminated. Failure to provide the ideal conditions in SSF can result in spoilage of fermented food.

Evolution of SSF Fermentors

Technologically, the SSF fermentors have not really evolved and are still stuck with the concept of racks or tray fermentors with little or no changes from the past. Many designs have been published that belong to the first category including static and agitated models but only few models are used in commercial production.

It is SF fermentors with its submerged fermentation that has undergone much technological evolution in size, design and function. This was really triggered by the birth of industrial microbiology spurred by the world wars.

Even today, bioprocess engineers are coming up with newer and unorthodox design for submerged fermentation in fermentors for the production of new metabolites from different type of cells. But seriously speaking the high yield of fermentation products such as penicillin is more due to the successful account of strain development rather than significant improvement in bioreactor design

This development of SF over SSF does not mean that SSF is irrelevant. It is more a reflection of the poor understanding and limitations of the various physical, chemical and biological processes to be controlled in any SSF process. It could also mean that there is a poor appreciation of the potential of SSF in the fermentation industries.

The main limitations of engineering design of SSF are:

a Problem of heat dissipation.

b Mixing.

The main method of heat dissipation is by evaporative cooling. In this process the loss of heat is correlated with loss of water from SSF. Thus it is crucial while there is a need to cool the heat there is also the need to retain the moisture for proper SSF to procee

While good mixing is important to obtain good mass transfers in SSF in drum type fermentors, the product of SSF might not be suitable as in the case of tempe fermentation where the final presentation of the tempe is important in itself.

Solid State Fermentation

In such fermentations, microbial growth and product formation occur at the surface of solid substrates. Examples of such fermentations are mushroom cultivation, mold ripened cheeses, starter cultures, etc. More recently, this approach has been used for the production of extracellular enzymes, certain valuable chemicals, fungal toxins, and fungal spores (biotransformation).

Traditional substrates are several agricultural products, rice, wheat, maize, soybean, etc. Such substrates selectively support mycelial organisms, which can grow at high nutrient concentrations and produce a variety of extracellular enzymes, e.g., a large number of filamentous fungi, and a few bacteria According to the physical state, solid state fermentations are divided into two groups:

(a) low moisture solids fermented without or with occasional/continuous agitation, and

(b) suspended solids fermented in packed columns, through which liquid is circulated. The fungi used for solid state fermentations are usually obligate aerobes.

Solid state fermentations on large scale use stationary or rotary trays. Temperature and humidity controlled air is circulated through the stacked solids. Less frequently, rotory drum type fermenters have been used.

Solid state fermentations offer certain unique advantages, but suffer from some important disadvantages. However, commercial application of this process for biochemical production is chiefly confined to Japan.

pH Control

INTRODUCTION

Most process plants generate a wastewater effluent that must be neutralized prior to discharge or reuse. pH control is needed in just about every process plant, and yet a large percentage of pH loops perform poorly. Results are inferior product quality, environmental pollution, and material waste. With ever increasing pressure to improve plant efficiency and tighter regulations in environmental protection, effective and continuous pH control is highly desirable.

The implementing a pH system is like putting a puzzle together. It will only work when all the components are in place. The pH puzzle includes effective pH probes, actuators, and controllers. While various pH probes and actuators for pH control are available, commercial adaptive pH controllers are still in demand. The challenge is to provide a controller that is able to deal with large non-linear gain changes in the pH loop. It will be useful for not only wastewater neutralization, but also chemical concentration control, since concentration is a key quality variable.

Temperature control

Temperature control is a process in which change of temperature of a space (objects collectively therewithin) is measured or otherwise detected, and the passage of heat energy into or out of the space is adjusted to achieve a desired average temperature.

CONTROL LOOPS

A home thermostat is an example of a closed control loop: It constantly assesses the current room temperature and controls a heater and/or air conditioner to increase or decrease the temperature according to user-defined setting. A simple thermostat merely switches the heater or air conditioner either on or off, and temporary overshoot and undershoot of the desired average temperature must be expected. A more expensive thermostat varies the amount of heat or cooling provided by the heater or cooler, depending on the difference between the required temperature (setpoint) and the actual temperature. This minimizes over/undershoot. The process is called PID and is implemented using a PID Controller.

ENERGY BALANCE

An object's or space's temperature increases when heat energy moves into it, increasing the average kinetic energy of its atoms, e.g., of things and air in a room. Heat energy leaving an object or space lowers its temperature. Heat flows from one place to another by one or more of three processes: conduction, convection and radiation. In conduction, energy is passed from one atom to another by direct contact. In convection, heat energy moves by conduction into some movable fluid (air or water) and the fluid moves from one place to another, carrying the heat with it. At some point the heat energy in the fluid is usually transferred to some other object by means conduction again. The movement of the fluid can be driven by negative-buoyancy, as when cooler air drops and thus upwardly displaces warmer air (natural convection), or by fans or pumps (forced convection). In radiation, the heated atoms make electromagnetic emissions absorbed by remote other atoms, whether nearby or at astonomical distance. For example, the Sun radiates heat as both invisible and visible electromagnetic energy. What we know as "light" is but a narrow region of the electromagnetic spectrum.

Design and working of a Bioreactor

A bioreactor may refer to any device or system that supports a biologically active environment. In one case, a bioreactor is a vessel in which is carried out a chemical process which involves organisms or biochemically active substances derived from such organisms. This process can either be aerobic or anaerobic. These bioreactors are commonly cylindrical, ranging in size from litres to cubic metres, and are often made of stainless steel.

There devices are being developed for use in tissue engineering. The bioreactor may be classified as batch fed batch or continuous. A continuous bioreactor is the chemestat. The large scale immobilised cell bioreactors are:-

1. moving media;

2. packed bed;
3. fibrous bed; and
4. membrane.

BIOREACTOR DESIGN

Bioreactor design is a relatively complex engineering task, which is studied in the discipline of biochemical engineering. Under optimum conditions, the microorganisms or cells are able to perform their desired function with 100 per cent rate of success. The bioreactor's environmental conditions like gas (i.e., air, oxygen, nitrogen, carbon dioxide) flow rates, temperature, pH and dissolved oxygen levels, and agitation speed/circulation rate need to be closely monitored and controlled.

A heat exchanger is needed to maintain the bioprocess at a constant temperature. Biological fermentation is a major source of heat, therefore in most cases bioreactors need refrigeration. They can be refrigerated with an external jacket or, for very large vessels, with internal coils.

In an aerobic process, optimal oxygen transfer is perhaps the most difficult task to accomplish. Oxygen is poorly soluble in water—even less in fermentation broths—and is relatively scarce in air (20.95 per cent). Oxygen transfer is usually helped by agitation, which is also needed to mix nutrients and to keep the fermentation homogeneous. There are, however, limits to the speed of agitation, due both to high power consumption (Electrical motor) and to the damage to organisms caused by excessive tip speed bio.

NASA has developed a new type of bioreactor that artificially grows tissue in cell cultures. NASA's tissue bioreactor can grow heart tissue, skeletal tissue, ligaments, cancer tissue for study, and other types of tissue.

NON-LINEAR DYNAMICS

It is a ridiculous term like "non-elephant zoology. To understand this phrase, imagine the time when programmable computers didn't exist. In those days it was impossible to solve a differential equation like the equation of motion of a pendulum driven by a periodic force. That is, the solution couldn't be expressed in terms of well-known functions like in the case of the linearized equation of motion. Non-linear equations of motion can be solved only in rare cases. For that reason, physicists tried to build their theories on linear differential equations because they are easier to solve. And indeed, the most successful theories (like electrodynamics and quantum mechanics) are based on linear differential equations. Other even older theories dealing with physical phenomenon closer to everyday experience, like fluid dynamics, were less successful because their dynamics is non-linear.

BASIC TERMS OF NON-LINEAR DYNAMICS

Non-linear dynamics is a language to talk about dynamical systems. Here, brief definitions are given for the basic terms of this language. All these terms will be illustrated at the pendulum.

Dynamical system

A part of the world which can be seen as a self-contained entity with some temporal behaviour. In non-linear dynamics, speaking about a dynamical system usually means to speak about an abstract mathematical system which is a model for such an entity. Mathematically, a dynamical system is defined by its state and by its dynamics. A pendulum is an example for a dynamical system. A number or a vector defining the state of the dynamical system uniquely.

For the undriven pendulum, the state is uniquely defined by the angle and the angular velocity := d/dt.. In the case of driving, the driving phase is also needed because the pendulum becomes a non-autonomous system. In spatially extended systems, the state is often a field. Mathematically spoken, fields are functions with space coordinates as independent variables. The velocity field of a fluid is a well-known example.

Phase space:

All possible states of the system. Each point in the phase space corresponds to a unique state. In the case of the undriven pendulum, the phase space has two dimensions whereas for driven pendula it has three. The dimension of the phase space is infinite in cases where the system state is defined by a field.

Dynamics or equation of motion:

The causal relation between the present state and the next state in the future. It is a deterministic rule which tells us what happens in the next time step. In the case of a continuous time, the time step is infinitesimally small. Thus, the equation of motion is a differential equation or a system of differential equations:

$du/dt = F(u)$,

where u is the state and t is the time variable. An example is the equation of motion of an undriven and undamped pendulum. In the case of a discrete time, the time steps are non-zero and the dynamics is a map:

$un+1 = F(un)$,

with the discrete time n. An example is the baker map. Note, that the corresponding physical time points tn do not necessarily occur equidistantly. Only the order has to be the same. That is, $n < m$ implies $tn < tm$.

The dynamics is linear if the causal relation between the present state and the next state is linear. Otherwise it is non-linear.

What happens if the next state isn't uniquely defined by the present one? In general, this is an indication that the phase space is not complete. Thus, there are important variables determining the state which had been forgotten. Of course, this is a crucial point while modelling a real-life systems. But beside this, there are two important classes of systems where the phase space is incomplete: The non-autonomuous systems and the stochastic systems. A non-autonomous systems has an equation of motion which depends explicitly on time. Thus, the dynamical rule governing the next state not only depends on the present state but also at the time it applies. A driven pendulum is an example of a non-autonomuous system. Fortunately, there is an easy way to make the phase space complete. Simply include the time into the definition of the state! Mathematically, this is done by introducing a new state variable. Its dynamics reads d/dt = 1, or n+1 = n, depending on whether time is continuous or discrete. For the periodically driven pendula, it is also natural to take the driving phase as the new state variable. Its equation of motion reads

d/dt = 2f,

where f is the driving frequency (the angular driving frequency is 2f).

In a stochastic system, the number and the nature of the variables necessary to complete the phase space is usually unknown. Therefore, the next state can not be deduced from the present one.

A non-wandering set may be stable or unstable. There are two types of stability, a weaker and a stronger one:

Lyapunov stability:

Every orbit starting in a neighbourhood of the non-wandering set remains in a neighbourhood.

Asymptotic stability:

In addition to the Lyapunov stability, every orbit in a neighbourhood approaches the non-wandering set asymptotically.

Thus, a non-wandering set is either asymptotically stable, marginally stable or unstable. Asymptotically stable non-wandering sets are also called attractors. The basin of attraction is the set of all initial states approaching the attractor in the long time limit.

There is a scheme called linear stability analysis: Assuming u0(t) is the orbit you want to check whether it is stable or not. It is asymptotically stable if any infinitesimal small perturbations u decays. Thus, make the ansatz u = u0+u, put it into the dynamics, i.e., du/dt = F(u) for a time-continuous system, un+1 = F(un) for a time-discrete system.

du/dt = DF(u0) u

and

un+1 = DF(u0) un,

These linear equations of motion are justified as long as the orbit doesn't go too far away from u0. For fixed points, the fundamental solutions are of the form exp(st)us and (s)nus, resp., where s and us (s = 1, 2,..., dimension of the phase space) are the eigenvalues and eigenvector, respectively, of the Jacobian DF(u0). The eigenvalues s are the roots of the characteristic polynomial det(DF(u0)-)=0. The fixed point u0 is asymptotically stable if all eigenvalues s are inside a stability area of the complex plane. This stability area is the half-plane left of the imaginary axis, whereas in the time-discrete case it is the unit circle around the origin. If at least one eigenvalue is outside the stability area the corresponding fundamental solution would increase exponentially. Thus, the fixed point will be unstable. If an eigenvalue is just on the border of stability, you can not decide by linear stability analysis whether the fixed point is unstable, asymptotically or marginally stable.

The notion spiral and node are inspired by the flow near the fixed point. A pair of conjugated complex eigenvalues lead to a spiral, whereas a node is causes by two real eigenvalues of the same sign. Real eigenvalues of different sign lead to a saddle. In general, a saddle is a fixed point where at least one eigenvalue has a positive real part but also at least one eigenvalue has a negative real part. Near a saddle, an orbit is usually attracted at first but repelled later on. There are points in the phase space which approach the fixed point for t . They form the stable manifold. The eigenspace for the eigenvalues with negative real part is tangential to the stable manifold. The unstable manifold are built by all points approaching the fixed point for t - . Saddles and their stable manifolds are usually the boundaries of a basin of attraction. The undriven pendulum has a saddle point (corresponding to the upside-down equilibrium position) and a spiral if < 20 or a node if > 20 (see the flow on the previous page).

The number of attractors in a non-linear dynamical system can change when a system parametre is changed. This change is called bifurcation. It is accompanied by a change of the stability of an attractor. In a bifurcation point, at least one eigenvalue of the Jacobian gets a zero real part. There are three generic types of so-called co-dimension-one bifurcations (the term codimension counts the number of control parametres for which fine tuning is necessary to get such a bifurcation):

In a stationary bifurcation, a single real eigenvalue crosses the boundary of stability. In the Pendulum Lab, stationary bifurcations of limit cycles occur quit often. They are stationary bifuractions of fixed points in a discrete dynamical system because limit cycles are fixed points in the Poincaré map.

Hopf bifurcations occur when a conjugated complex pair crosses the boundary of stability. In the time-continuous case, a limit cycle bifurcates. It has an angular frequency which is given by the imaginary part of the crossing pair. In the discrete case, the bifurcating orbit is generally quasiperiodic, except that the argument of the crossing pair times an integer gives just 2. In the Pendulum Lab, Hopf bifurcations seem to be absent.

In discrete dynamical systems, a certain bifurcation is possible which is absent in continuous systems: The period doubling bifurcation. Here, a real eigenvalue crosses the boundary of stability at -1. In a period-doubling bifurcation point, a period-2 orbit bifurcates. You can find this type of bifurcation quite often in the Pendulum Lab.

Crossing the boundary of stability only indicates a bifurcation point and the type of the bifurcating solutions. But it doesn't tell you how and how many new solutions bifurcate or disappear in a bifurcation point. To answer this question one has to take into account the leading non-linear terms. This might appear to be a difficult task. But fortunately, a theorem and a transformation help us simplifying this analysis:

Centre manifold theorem:

It helps to reduce the dimensionality of the phase space to the dimensionality of the so-called centre manifold which in the bifurcation point is tangentially to the eigenspace of the marginal modes of the linear stability analysis (i.e., the eigenfunctions which neither decay nor expand exponentially). The centre manifold theorem says, the dynamics can be projected onto the centre manifold without loosing any significant aspect of the dynamics. Thus, the dynamics near a stationary co-dimension-one bifurcation can be decribed by an effective dynamics in a one-dimensional subspace. Hopf-bifurcations need a two-dimensional phase space. This is exactly the minimum dimension for limit cycles in continuous dynamical systems and for quasiperiodic orbits in discrete dynamical systems. Period-doubling bifurcations need only a one-dimensional phase space.

Transformation to normal form:

The dynamics projected onto the centre manifold can by transformed to so-called normal forms by a non-linear transformation of the phase space variables.

The main normal forms (for continuous dynamical systems) of stationary co-dimension-one bifurcations are shown. The phase space variable is u. The control parametre is.

The bifurcation point is at = 0. The direction of motion in the one-dimensional phase space is shown by arrows. Stable (unstable) fixed points are drawn as red solid (dotted) lines. The normal forms are shown in blue.

The generic case of stationary co-dimension-one bifurcation is a saddle-node bifurcation. It is generic (that is, the probability is finite to pick a dynamical system having a saddle-node bifurcation) because fixed points lie on a smooth one-dimensional manifold in the combined space of phase space and control parametre. The minima and maxima of as a function of the curve length denote saddle-node bifurcations where a stable fixed point (a node) annihilates with an unstable one (a saddle in more than one dimension). A combination of a minimum and a maximum lead to the phenomenon of bistability where in a certain interval of the control parametre, two stable attractors exist with an unstable one in-between. Bistability is responsible for hysteresis in many physical and technical systems, as for example the button on your computer mouse. This phenomenon occurs if one sweeps the control parametre relatively slowly beyond the interval of bistability. When the interval of bistability is left the attractor disappears in a saddle-node bifurcation, and the system suddenly jumps to another attractor. For the example of a mouse button, this means that the button makes click even if pressing the button extremely slowly. Now sweeping the control parametre in the other direction has no effect when the bistability interval is re-entered. Thus, the mouse button remains in the pressed state even if the pressure is diminished below the pressure necessary for the previous click. Leaving the bistability interval at its other end, the system has to jump towards the former attractor.

And the mouse button makes again click. In the Pendulum Lab, bistability occurs e.g. in the foldover effect.

The transcritical bifurcation occurs when in the combined space of phase space and controlparametre space two different manifolds of fixed points cross each other. At the crossing point the fixed points exchange there stability property. That is, the unstable fixed point becomes stable and vice versa. Note, that beyond the bifurcation point the number of fixed points has not changed contrary to saddle-node bifurcation where two fixed points appear or disappear. A transcritical bifurcation is not a generic bifurcation in a phase space with more than one dimension because it is unlikely that two lines cross each other in a space with more than two dimensions.

Pitchfork bifurcations are possible in dynamical systems with an inversion or reflection symmetry. That is, an equation of motion that remains unchanged if one changes the sign of all phase space variables (or at least for one). An example is the undriven pendulum. Usually, such a system has a symmetric fixed point (or limit cycle). Pitchfork bifurcations are the generic bifurcations when such a symmetric solution changes its stability. The bifurcating solution does not have the full symmetry of the equation of motion. This phenomenon is called broken symmetry. A solution with a broken symmetry does not occur in isolation because the broken symmetry applied onto such a solution generates a new solution where the same symmetry is broken. All such solutions build a family. Therefore, always two fixed points with a broken symmetry bifurcate at once in a pitchfork bifurcation. Both are either stable (supercritical pitchfork bifurcation) or unstable (subcritical pitchfork bifurcation).

In the Pendulum Lab, pitchfork bifurcations of limit cycles occur quite often. As an example, try the following experiment: Shake a horizontally driven pendulum faster than its eigen frequency. It will show a tiny oscillation around the equilibrium value = 0. This limit cycle becomes unstable if the driving frequency is much larger than the eigenfrequency. At the bifurcation point, two new stable limit cycles bifurcate supercritically. Their centres are on the right and on the left hand of the unstable symmetric limit cycle.

The bifurcation diagrams of a Hopf and a period doubling bifurcation are similar to the diagram of a pitchfork bifurcation. That is, the bifurcating periodic or quasiperiodic solution is either stable (supercritical bifurcation) or unstable (subcritical bifurcation). Again, a broken symmetry is responsible for this similarity. Here, it is the invariance of the dynamical system against translations in time.

THE NATURE OF DETERMINISTIC CHAOS

While you played with the driven pendula in the lab, you may have discovered that it can behave very erratic. For example, you may have noticed irregular sequences of left and right turns. This behaviour, called deterministic chaos, is the most prominent effect of non-linear dynamics. The term "deterministic chaos" seems to be a contradiction in itself. How can something be deterministic and chaotic at the same time?

1. Sensitive dependence on the initial conditions

g = 9.81 m/sec2, l = 1 m, = 1/sec, a = 0.85 m, f = 1 Hz.

A key element of deterministic chaos is the sensitive dependence of the trajectory on the initial conditions. As an example, the plot on the left shows the first 15 seconds of a demonstration of this effect for a horizontally driven pendulum. The initial conditions are

pendulum A: = -140°, d/dt = 0

pendulum B: = -140°1', d/dt = 0

Both initial conditions differ in one arcsec only. Nevertheless, 10 seconds later they behave totally different. At the beginning the distance of the trajectories increases on average exponentially (see the inset which shows the distance in phase space). The rate of divergence is measured by the largest Lyapunov exponent.

How predictable is a chaotic system? As any deterministic system, it is predictable on a short time scale. But on a long time scale, chaotic systems become unpredictable. They behave seemingly irregular. It is this very behaviour why such systems are called "chaotic". Predictability is lost at a time, called prediction horizon, that depends only logarithmically on the uncertaincy of the initial conditions. Thus, the prediction horizon increases fairly slowly with increasing knowledge of the initial condition. There is an ultimate prediction horizon because the error of the initial condition can not be smaller than the unavoidable thermal noise. We can not foresee the future beyond this horizon! It depends on the system how far away this horizon actually is. For the example shown in the plot, it is about 8 seconds. For the weather, it may be roughly 10 days (maybe in places like southern California it can be quite longer). And for the our solar system with all planets, the prediction horizon lies very far in the future, namely, a few million years (luckily, otherwise astronomy wouldn't be the paradigma of successful predictions).

2. Stretching and folding

How can we understand deterministic chaos? Note, that sensitivity on the initial conditions isn't a privilege of non-linear dynamics. It also happens in linear systems! For example, take the following time-discrete dynamical system

xn+1 = 2xn.

How can we understand deterministic chaos? Note, that sensitivity on the initial conditions isn't a privilege of non-linear dynamics. It also happens in linear systems! For example, take the following time-discrete dynamical system

xn+1 = 2xn.

The distance between two different solutions increases by the factor two in each time step. But of course, this is not deterministic chaos! It is a kind of explosion (like the population explosion!). Sensitivity on the initial conditions leads to chaos only if the trajectories are bound. That is, the system can not blow up to infinity. With linear dynamics, you can have either sensitivity on the initial conditions or bound trajectories, but not both. With non-linearities, you can have both. The figure shows why this is possible. Imagine the phase space is a piece of dough from which you want to make flaky pastry for croissants: You roll it out and fold it. Usually you repeat this step a few times. Suppose at the beginning two poppy-seeds are located next to each other. Each time the dough is rolled out the distance between the seeds increases. Eventually, this increase is stopped when the seeds are on different sides of the folding line. At that moment the distance changes unpredictable. Assuming an initial distance of 1 mm and a piece of dough which is 10 cm long, you reach the prediction horizon after only seven rolling-folding steps. Thus, stretching and folding are responsible for deterministic chaos. And there is no folding without non-linearities!

3. The baker map

There is a simple map on a two-dimensional phase space, called baker map, which is an abstract version of making flaky pastry:

$x_{n+1} = 2x_n \bmod 1, \quad y_{n+1} = y_n/2 + (x_n - 1/2)/2,$

where is the step function which is 1 for positive arguments and zero otherwise. The baker map is a piecewise linear map of the unity square onto itself. First, the square is stretch in the x direction by the factor two and squeezed in the y direction also by the factor two. In the second step, the stretched and squeezed square is cut in the middle and the right half is put onto the left half. This corresponds to the folding step. Note, that the map preserves the size of an area during mapping. The sequence of pictures shows the mixing properties of the baker map when it is applied to the logo of the pendulum lab.

4. Kneading the phase space of the pendulum

This animation of a horizontal pendulum (with the same parametres as above) shows how the phase space is stretched and fold during one cycle of driving. Contrary to the baker map the phase space shrinks because of dissipation. This animation was produced by calculating the orbits of 1600 initial conditions which are distributed uniformly over the visible phase space. After a transient the initial rectangle shrinks neither to a point nor a line but to something which is called a strange attractor. It has a structure similar to flaky pastry. After each cycle the number of leaves is increased by some factor. Thus, in the longtime limit the strange attractor has infinitely many leaves. Nevertheless, the volume in phase space is zero! Such a geometric object is indeed strange. But nowadays, it is well-known as fractals.

9

BIO-FERMENTATION TECHNOLOGY

Bio-fermentation technology makes it possible to grow a number of fungi in large tanks and, in a matter of days, a large quantity can be produced. The technology has now become highly advanced and is making previously rare herbs like Cordyceps and Ganoderma much more accessible. Many studies indicate that the chemical nature of this biotechnology Cordyceps is almost identical to that of the wild variety. The Cordyceps contained in most commercial products is produced by this technology.

ADVANTAGES OF BIO-FERMENTATION

Standardization

Highly controlled by scientific means during its growth and "standardized", every batch is virtually identical and, in a sense, "perfect." This is very important in pharmaceutical terms because without standardization, it is difficult or impossible to develop drug-type standards for substances like Cordyceps. Once a herb can be standardized, all kinds of studies can be conducted that will be accepted by the scientific community. However, standardization, by itself, does not make a product effective.

Cost

Products can be produced more cheaply by bio-fermentation. High quality wild Cordyceps is rare and now sells for well over a thousand dollars a pound in cities like Hong Kong, Tokyo, New York or London. The culture-grown Cordyceps is available at a third of the cost, with approximately the same benefits. Fermentation technology makes this substance more readily available.

Vegetarian

Wild Cordyceps, by weight, is mostly caterpillar. For vegetarians, and those who do not want to consume caterpillars, the new technology provides a solution. The fermentation technology does not include the caterpillar in the growing process. The fungus is grown without the use of animal nutrients and the result is a 100% pure "vegetarian" health product.

Wild Cordyceps

Studies show that wild Cordyceps is richer in certain components and the proportion of components is different, which probably makes a difference in activity. Cultivated Cordyceps is still a premium tonic herb and the Cordyceps most commonly found in Chinese herb stores is cultivated.

Supply of bio-fermentation Technology

Fermentation technology to make crop stalks and straw waste into a green pollution-free bio-feed for livestock breeding, animal manure and other waste but also rural and urban solid waste by biological agents quickly look into a high bio-organic fertilizer into the green plant. Agent can look to a variety of fermentation of crop straw, melon vine, animal manure, leaves, weeds, and bran residue of vinegar, wine dregs, residue of vinegar, soy sauce residue, soybean, mushroom residue, powder residue, rubbish, bone meal, bagasse and other waste into organic fertilizer and bio-fast feed.

Fermentation of organic fertilizer and a ton of finished feed, just use 2 kg of bio-bio complex microbial agents, at least a week, as many as two weeks, all green waste can be turned into pollution-free eco-fertilizers and bio feed. The fertilizer and feed production is simple and low cost. Use of their houses in rural areas, idle houses and fields, can be produced throughout the year to reduce the human transport, local production, local use. Is the majority of farmers do not buy fertilizer and feed friends, turning waste into the wealth channel.

Fertilizer complex microbial agents 18 yuan per kg, straw fermentation agents 30 yuan per kg, 25 kg / bag from the mail, free shipping over 1 ton and 10 ton (where the joint venture construction of the user), the production company free of charge technical services. Look biotechnology is organic fertilizer plant, feed mill, factory farming, sewage treatment plant, township, town, village, ideal for entrepreneurial farmers.

INDUSTRIAL PRODUCTION OF ETHANOL

History of Ethanol Production:

- Louis Pasteur demonstrated the fermentation of sugar by microbes and their regulations as well. He found that the yeast cells produce 20 times more cell materials under aerobic conditions than under unaerobic conditions. He explained that Oxygen inhibits fermentation process. Yeasts are aerobic microorganism, but glucose fermentation takes place under unaerobic condition. This is also known as" Pasteur effect ".
- In 1815, Gay- Lussac formulated the conversion of glucose to ethanol.
- During 1980, lactic acid was produced by fermentation as a substitute for tartaric acid used in baking powder which was not commercialized.

Characteristics of Ethanol:

- Ethanol is a volatile, flammable, colourless liquid.
- Ethanol is a straight-chain alcohol, and its molecular formula is $C2H5OH$.
- Its empirical formula is $C2H6O$, making it a constitutional isomer of dimethyl ether.
- The fermentation of sugar into ethanol is one of the earliest organic reactions employed by humanity

Production:

The entire process of ethanol production is divided into two stages namely upstream process and downstream process.

Upstream Processing:

Upstream processing encompasses any technology that leads to the synthesis of a product as well as the fundamental science and engineering needed to understand product formation. Upstream includes the exploration, development and production of crude oil, bitumen and natural gas etc.

Types of Ethanol Production:

Ethanol is produced both as a petrochemical, through the hydration of ethylene, and biologically, by fermenting sugars with yeast.

Ethanol Hydration:

Ethanol hydration is most often made from petrochemical feed stocks, typically by the acid-catalyzed hydration of ethylene. The catalyst is most commonly phosphoric acid absorbed on to a porous support such as diatomaceous earth or charcoal. This catalyst was first used for large scale production by shell Oil Company in 1947. The reaction is carried out with an excess of high pressure steam at 300°C.

Microorganisms Used in Ethanol Production:

Some of the microorganisms that are used in the fermentation process are mentioned below.

Bacteria:

Clostridium acetobutylicum; Klebsiella pneumoniae; Leuconostoc mesenteroides etc.

Fungi:

Aspergillus oryza; Mucor sp; Neurospora crassa; Saccharomyces sp. (yeast); etc

Selective Features for The Strains:

- It should be an efficient strain. In otherwise it should produce a large quantity of alcohol.
- It should be fast growing.
- It should have a high tolerance to alcohol as well as to osmotic pressure.
- It should posses uniform and stable biochemical properties.

Preparation of Inoculum:

After selection of the desired organism and its isolation in pure form, the inoculum is prepared under aseptic conditions. For this purpose the organism are first cultured in flasks under aerobic conditions to increase the size of the inoculum which can be used for inoculation.

RAW MATERIALS:

A variety of carbohydrates available as waste products of agricultural industries are used for the preparation of fermentation medium on the basis of their chemical nature. They are grouped as;

- Saccharide materials : molasses, whey, glucose, sucrose.
- Starchy materials : potato, wheat, rice, corn, starch.
- Cellulosic material : waste liquor, wood.

Pre-treatment of Raw Materials:

Most of the raw materials of alcohol fermentation require some degree of pre treatment. The actual process depends on the chemical composition of the raw materials. In general, the sugary raw materials require mild or no pretreatment, while the cellulosic materials need extensive pretreatment. This is because the cellulosic substances have to be subjected to acidic or enzyme hydrolysis to release monosaccharide units that are needed for alcohol production.

Fermentation Conditions

Carbon Sources:

Sugar concentration in the range of 10 to 18 % is found satisfactory. Sugar concentration of 12 % is often used. For example, cane molasses contain sucrose as a carbon source. High concentration affects yeast adversely while low concentration makes the process uneconomical.

Nitrogen Sources:

Many organic as well as inorganic nitrogenous compounds may be used. Ammonium sulphate is generally used. The level of nitrogen in the medium is limited.

Growth Factors:

It is not necessary to add the growth factors separately in the preparation of the fermentation medium. Since the raw materials used satisfy the requirements of the culture.

PH:

PH of the fermentation medium is 4.8 to 5. The higher PH increases the chances of contamination, whereas lower PH values inhibit the yield of ethyl alcohol.

Temperature:

Usually, the temperature range of 70 to 80°F is preferred. Temperature can be controlled by cold water spray around the fermentation tank. Higher temperature favours the growth of bacteria and also causes loss of alcohol due to evaporation.

Fermentation Method

Fermentation of ethanol is carried out in large fermentation tanks. The Inoculum of microorganisms is maintained in the fermenter at the optimum growth conditions. At the same time fermentation medium is formulated and transferred to the sterile fermenter. Liquid medium in fermenter is inoculated with a small amount of inoculum of yeast. Fermentation starts within a few hours after the addition of yeast. The process becomes rapid after 24 hours to complete. Duration of ethyl alcohol fermentation is usually 30 to 72 hours when the specific gravity of the fermented liquid becomes constant.

Downstream Processing

The extraction and purification of a biotechnological product from fermentation is referred to as downstream processing.

STAGES IN DOWNSTREAM PROCESS

Solid Liquid Seperation

The first step in product recovery is the separation of whole cells and other insoluble ingredients from the culture broth. Some of these methods are;

Floatation

When a gas is introduced into the liquid broth, it forms bubbles. The cells and other solid particles get adsorbed on to these gas bubbles. These bubbles rise to form foam layer which can be collected and removed.

Floculation

In this, the cells form large aggregates to settle down for easy removal. The process of flocculation depends on the nature of cells and ionic constituents of the medium.

CONCENTRATION

Liquid-liquid Extraction

The concentration of biological products can be achieved by transferring the desired product from one liquid phase to another liquid phase, which is referred to as liquid-liquid extraction.

Adsorption

The biological product of fermentation can be concentrated by using solid adsorbent particles.

Purification

The biological products of fermentation are very effectively purified by chromatograph techniques. One such is;

Ion-exchange-exchange Chromatography

It involves the separation of molecules based on their surface charges. Ion exchangers are of two types- cation exchangers with negative charged groups and anion exchangers with positively charged groups. The most commonly used cation-exchangers are dowex HCR and amberlite IR, the anion- exchangers are dowex SAR and amberlite IRA.

In ion-exchange chromatography, the PH of the medium is very crucial, since the net charge varies with PH. In other words, the PH determines the effective charge on both the target molecules and the ion exchanger. The ionic bound molecules can be eluted from the matrix by changing the PH of the eluant or by increasing the concentration of salt solution. Ion-exchange chromatography is useful for the purification of antibiotics, besides the purification of proteins.

Product Recovery

Ethanol is separated from the fermentation broth in continuous stilts resembling those utilized for the recovery of acetone, butanol, and ethanol of 95% concentration is obtained by successive distillations in these stilts. To obtain alcohol concentrations greater than 95%, however requires special distillation techniques because of the ability of alcohol to form an azeotropic mixture of benzene, water and ethanol, which is then distilled with increasing temperature increaments. This procedure removes first the azeotropic benzene-water-ethanol mixture, and then an ethanol-benzene azeotropic mixture so that absolute alcohol remains. The absolute alcohol as well as 95% ethanol is marketed with and without denaturation.

A high-boiling fusel oil fraction also is recovered during distillation of the fermentation broth and this fraction accounts for approximately 0.5% of the crude distillates. Other products of the fermentation, except for the carbon dioxide which is often collected for reuse are present in only small amounts. These trace amounts of succinic acid and slightly larger amounts of glycerol occur. Some acrolein also may be present in the distillate, have been formed by microbiological contaminants attacking the glycerol.

BIO-FERMENTATION TECHNOLOGY TO IMPROVE EFFICIENCY OF SWINE NUTRITION

In ancient times, mankind survived with food from daily hunting. As human being accumulates experiences and knowledge about nature, agriculture became a major way of obtaining food. Agricultural activity also allowed us to have stable food supply and this was one of important reasons for the increase in human population and cultural flourish. As entering modern society, growth of human population accelerated as so demands for food. Limited space and resources with fast growing demands for food has pushed agriculturists to improve efficiency of food production through agriculture. Animal agriculture (or livestock production) is an important part of agriculture producing meat and fat. Pork production takes a key role providing over 42% of global meat consumption.

Swine Nutrition and Biotechnology

We can't ignore that continuous genetic selection has made great contribution to animal agriculture by improving genetic potentials of maximal productivity. Scientific advances in nutrition, then, allowed expressing these genetic potentials to maximize productivity of livestock. It is amazing to see the improvement of productivity of pork production during the 20-40 years. The author had a chance to compare few economically important traits of pigs (especially for sows) between 1979 and 1999 (Kim, 1999).

During the past 20 years, performance of sows has been improved dramatically. As examples, litter size of sows has increased by 9% whereas litter mortality has reduced by 18%.

It is not difficult to see the evidences showing that biotechnology was a key driving factor for the great improvement in the efficiency of pork production. One can easily think genetic engineering as a main theme of biotechnology. However, genetic engineering is only a portion of biotechnology. United Nations Convention on Biological Diversity defines biotechnology as "Any technological application that uses biological systems, dead organisms, or derivatives thereof, to make or modify products or processes for specific use". Biotechnology indeed includes a broad range of applications including bioengineering, bio-remediation, gene therapy, genetic selection, bio-degradation, cloning, bio-fermentation, etc.

Among these, we can find many evidences that pork production has much been benefited from application of bio-fermentation technology in preparing feed and feed additives. In the next sections, few examples of successful application of bio-fermentation technology in swine nutrition are described based on the author's own research experiences.

Supplemental Amino Acids

It is early 1960s when an amino acid, mono-sodium glutamate, was produced in a large scale using biofermentation technology to be used for food application. Soon after, similar bio-fermentation technology was applied to produce feed grade supplemental amino acids.

Supplemental amino acids are produced by liquid state fermentation. Briefly, sugars from sugar cane are extracted and used as a substrate for fermentation by microorganisms specifically selected for synthesizing targeting amino acids. After these microorganisms excrete targeting amino acids to broth, drying process and filtration are followed to obtain crystalline amino acids. Good examples of supplemental amino acids that can be produced in large scales are Llysine HCl, L-threonine, and L-tryptophan which are, in fact, three key limiting essential amino acids for swine. In recent years, feed grade supplemental L-glutamine and Lvaline are also introduced by a company leading this area. L-arginine HCl is another good example of supplemental amino acid for its unique function in an animal body that is also produced by bio-fermentation.

Production efficiency of supplemental amino acids through bio-fermentation has been improved significantly due to advances in biotechnology. Currently production efficiency of L-lysine reaches over 50%.

L-lysine HCl is the first feed grade supplemental amino acid that is produced in a large scale by bio-fermentation which is soon after production of mono-sodium glutamate for food application. It was early 1960s when feed grade Llysine HCl was mass-produced first time in Japan. Since then L-lysine has well been used as the first limiting essential amino acid in swine feed. It may be 1950 when the first research paper about the requirement of lysine for swine published in a scientific journal (Brinegar et al., 1950).

Production efficiency of lean pork has been maximized as nutritionists can formulate swine feed matching requirements of these limiting essential amino acids and reducing crude protein contents. It can be formulated to include 11.7% crude protein with the use of L-lysine, Lthreonine, and L-tryptophan which would provide an equal nutritional value of feed containing 15.2% crude protein without supplemental amino acids.

Recently, a large scale production of feed grade L-valine has been initiated using bio-fermentation technology. Use of L-valine, which is often 4th or 5th limiting essential amino acids for swine, allows nutritionists to further reduce crude protein content in the feed without affecting growth performance of pigs. When these four feed grade amino acids are used in swine feed, 1% value of crude protein content can be further reduced. Currently feed grade Lvaline is only marketed in European Union, however, the use of L-valine will soon be extended to Americas and Asian countries. Development of new feed grade L-valine allowed nutritionists to formulate swine feed with high accuracy for maximal lean growth as well as that is environmentally friendly by reducing contents of excess amino acids in the diets over the requirements.

Another feed grade amino acid that is recently development by bio-fermentation is L-glutamine. Concept of applying L-glutamine in swine feed is somewhat different from the use of previous four limiting amino acids.

L-glutamine is known as a non-essential amino acid that is not a limiting amino acid for swine. However, recent research clearly indicates specific function of L-glutamine other than being used for protein synthesis in an animal body. It is demonstrated that L-glutamine is a major energy substrate and an essential precursor of Nacetylglucosamine-6-phosphate for proliferation and growth of enterocytes and for prevention of intestinal atrophy. L-glutamine is essential for development of gut function especially for newly weaned young piglets that are under stressful condition due to new environment and diet changes. It is also suggested that pigs under chronic or acute stress with immune depression can be benefitted by supplemental Lglutamine as immune cells use L-glutamine as a major energy source.

L-arginine is another important example of an amino acid that has great potential to be used in swine feed even though a large scale feed grade production has not occurred yet. There are many research papers showing unique functions of L-arginine in an animal body including humans.

Briefly, L-arginine can be used by endothelial cells lining blood vessels which possess an enzyme called nitric oxide synthase converting L-arginine to citrulline releasing nitric oxide. Production of nitric oxide causes conversion of cellular GTP to cGMP which will cause vasodilatation with increase blood flow. In the mean time, L-arginine can also be converted to orinithine which will in turn metabolized to a group of polyamines including spermine and spermidine.

Polyamine is known as a key components used for angiogenesis and cell proliferation. A recent study showed that dietary supplementation of L-arginine to milk replacer improved growth of newly born piglets. Sows are also benefitted by dietary supplementation of L-arginine. Improved blood flow to placenta and mammary gland increased nutrient supply to fetuses for growth and to mammary gland for milk production. Sows with supplemental L-arginine had reduced number of dead born piglets and increased milk production during early stage of lactation.

ENZYMES

Rapid increase in the cost of feed ingredients caused such a big obstacle to swine producers. Shortage of good quality feed grains pushed swine producers to include grain sources that are not in a best quality with indigestible nutrients and with potential anti-nutritional factors and mycotoxins. This caused decrease in production efficiency due to reduced utilization of nutrients in the feed by pigs. Use of exogenous microbial enzymes has greatly been increased in feeding swine. Most of exogenous enzymes are produced by bio-fermentation to help digestion of specific nutrients or compounds in the feed or feed grains. Thus specific enzymes can be supplemented to a diet with specific ingredients with potential problems. One of popular examples is the use of phytase to hydrolyze phytic acid structure releasing phosphorus as well as other minerals.

There are another types of exogenous enzymes that became popular. Carbohydrases or more often called NSPases (non-starch polysaccharidases) helps efficiency of energy utilization of feed grains by pigs. Application of exogenous NSPases should consider types of feed grains used in the diet because types of NSPs are different depending on feed ingredients (Kim and Baker, 2003). For corn-soybean meal based feed that are most widely used in Americas and Asian countries, types of NSPs that interfere energy digestibility are galactosides, mannosides, soybean oligosaccharides (raffinose, stachyose, and verbascose).

Growing-finishing pigs can also be benefitted by the use of NSPases as their ability to utilize energy in the feed improves. Swine producers can reduce energy content in the diets without sacrificing growth performance when exogenous NSPases are supplemented. Use of NSPases in lactation diets helped sows often possessing a limited voluntary nutrient intake to preserve body tissues without reducing milk production as their energy digestibility improved.

As previously described, exogenous microbial enzymes produced by bio-fermentation has made a great contribution to swine production by improving production efficiency.

Yeast Metabolites

There have been many types of probiotics and prebiotics introduced as potential alternatives to the use of supplemental antibiotics as growth promoters. These probiotics and prebiotics mainly target reducing pH and/or reducing population of pathogenic microorganisms in hind

gut. However, lack of consistent research evidences showing benefits of probiotics and prebiotics limits application in swine diets. Yeast metabolites have newly been introduced in swine diets. Yeast metabolites are also called yeast culture that includes concepts of both probiotics and prebiotics. Yeast metabolites are first introduced to dairy producers and known to improve dry matter intake of dairy cows increasing milk production (Putnam et al., 1997).

Recent research with sows conducted at a large commercial swine operation introduced that yeast metabolites may increase milk production for litter growth.

Yeast metabolites are produced by bio-fermentation of cereal grains by Saccharomyces ceravisiae under anaerobic liquid state condition. Yeast cells would preserve essential metabolites for their survival instead of generating ATPs under an anaerobic condition. Solid state fermentation follows liquid state fermentation encouraging further production of metabolites that can be dried and used in feed application. Yeast metabolites, therefore, include metabolites produced during anaerobic fermentation as well as cell content and cell wall components. Repeatable results are obtained showing dietary supplementation of yeast metabolites improves milk production in sows, even though it is premature to conclude which component in yeast metabolites helps milk production at this moment.

Fermented Ingredients

Vegetable protein sources including soybean meals have economical advantages over expensive animal protein sources. However, these vegetable protein sources often contain anti-nutritional factors limiting their application in feeding newly weaned pigs with immature gut structure that can be damaged by anti-nutritional factors. There are

several anti-nutritional factors found in vegetable protein sources including trypsin inhibitor, lectin, flatulence producing compounds, phytate, glycinin, â-conglycinin, etc.

Some of anti-nutritional factors can be destroyed by feed processing such as heating and cooking. Newly weaned pigs with limited HCl in stomach and enzymatic secretion in small intestine can have difficulties to digest proteins with complex structures and large molecular weights.

There has been an effort to improve vegetable protein sources by reducing anti-nutritional factors using bio-fermentation technology. Aspergillus oryzae and Bacillus subtilis have traditionally been used preparing soybeans via fermentation. Careful selection of these microorganisms for high activity in specific feed ingredient allowed preparation of novel fermented vegetable protein sources. This fermentation process by Aspergillus oryzae and Bacillus subtilis reduced trypsin inhibitor below a detection limit, reduced phytate content by 4-folds, reduced size of peptides for suitable for digestion, and reduced contents of glycinin and â-conglycinin that are known as allergenic compounds in gut.

Use of these fermented vegetable protein sources in nursery diets improved growth efficiency of newly weaned pigs compared to pigs fed same vegetable protein sources without fermentation. Fermented vegetable protein sources could also replace the use of animal protein sources such as whey protein concentrates and spray dried plasma protein in part without compromising growth performance. It is a great contribution to swine industry applying bio-fermentation technology to prepare relatively inexpensive vegetable protein sources replacing the use of expensive animal protein sources in swine diets.

The Biofermentatio Technology allows for 100X of effective product to be dosed in to a specific environment thereby providing astonishing results in turning around biological wastewater treatment systems in days NOT WEEKS, starting up plants in hours NOT WEEKS.

ABS is an Enviromental Technology Company with over 20 years experience and 50+ patents. ABS has a proven commercialized technology called Biofermentation™ that can grow thousands of pounds of oil eating microbes at a "Staging Facility" in your county or city, which, when applied, result in "Accelerated Natural Attenuation." This technology has been given a favorable review by BP Horizon Spill Response. The technology has been presented on CNN by an ABS distributor RTG (Todd Brown). Biofermentation™ offers the only solution to scale-up the oil eating microbes necessary to bioremediate marshes and the oil stains left on the beaches after removal of tar balls.

What are Microbes?

Microbes are living organisms such as bacteria or fungi. ABS ONLY uses NATURAL STRAINS OF MICROBES, genetically engineered microbes (GEMS) are highly regulated and are not legally allowed to be dispersed in to the environment without specific controls.

What do Microbes Do?

Microbes eat organic compounds to grow and survive converting these organics to harmless end products of carbon dioxide, water and more microbes. Many microbes are useful for biodegrading organic contamination, while others use hydrogen sulfide or ammonia. The latter microbes can be harnessed to control odors or to enhance nitrification. The process of biodegradation involves a complex series of reactions involving many enzymes, so microbes are essentially enzyme factories, each enzyme having a specific purpose, but with the microbe managing the ultimate outcome.

ADVANTAGES AND DISADVANTAGES OF FERMENTATION TECHNOLOGY

Aqua Culture is the fermentation bed utilization microbiology, ecology, fermentation Engineering principles to microbial activity as a function of material and energy "conversion center," an ecological farming model. Mainly in the barn covered with a layer of the a certain thickness and species of the sawdust litter, bacteria and the use of animal feces and urine discharged from the decomposition of sawdust as a source of nutrition and transformation, restored the arch of pigs eating habits, some bacterial proteins can be pig feed, the whole External rearing process to achieve zero emissions, no odor, no pollution of the effect of making long-term function of micro-organisms, sustainable and stable play effect. Fermentation Technology is being pig-promotion organizations and business organizations a number of widely publicized, although the fermentation pig technology has many advantages, but in practice some problems can not be ignored, now pig-technology advantages and disadvantages of fermentation are described below for reference peers.

Pig-1 fermentation technology advantages

No Emissions, no odor, no pollution

Pig-fermentation technology is different from the traditional pig technology, pig, pig urinary retention in the long term, not out emissions, not to the surrounding flow, relying on the role of microbial decomposition, transformation; and unique free pig retractable roof design, ventilation and heat transfer, so that the formation of cold and warm air convection, hot air flow, while the bottom of the circle at the end of the water evaporates, the bottom circle will remain the best condition, no smell, and when feces and urine into the feed resources after the pig feed, not removed, will be less and less. adequately address the technology of traditional water resources caused by pig waste and environmental pollution problems.

Save feed and reduce costs, improve efficiency

Because manure, urine is broken down by microorganisms into inorganic substances can be pigs and eat bacteria proteins, and sawdust in the wood fiber and semi-fibers can also be degraded into readily fermentable sugars, turned arch through the consumption of pork, to the Pigs provide some protein and other nutrients, thereby reducing the supply of concentrate feed. According to reports, this bed is not only fermented swine never have to clean up litter bed is the fermentation, 10 to 20 years do not have to change. In addition, because pig-way without artificial fermentation

Conservancy, washing pig bed, cleaning pens, feeding the one hand, to reduce staff to save labor costs, on the other hand saves water.

10

FERMENTERS AND BIOREACTORS

A Fermenter is a container or vessel of food grade quality, in which yeast ferments fruit or grain. A fermenter can be any thing from a 5 gallon bucket to a multi thousand gallon vessel. But for the Home Distiller any thing from 5 gallons to 15 gallons is more than enough do to the fact that our stills are no larger than a 15 gallon boiler. Fermenters usually have an airlock which allows carbon dioxide to escape, but prevents air from entering the container. In an enviroment rich in oxygen, the yeast will use aerobic respiration which does not produce alcohol.

DIFFERENT TYPES OF FERMENTERS

Bioreactors

A bioreactor is a device in which the microorganisms are cultivated and motivated to form the desired products by maintaining optimum conditions for growth and metabolic activity. A fermenter refers to the device used for the cultivation of prokaryotic cells e.g. bacteria, fungi etc. whereas a bioreactor is used for growing eukaryotic cells. A typical conventional bioreactor has cylindrical vessel with domed top and bottom generally made up of stainless steel. The reaction vessel which is surrounded by a jacket, is provided with a sparger at the bottom through which air (or other gases such as CO2 and NH3 (for pH maintenance) can be introduced. The reaction vessel also has side ports for pH, temperature and dissolved O2 sensors. The agitator shaft is connected to a motor at the bottom. Above the liquid level of the reaction vessel, connections for acid, alkali, antifoam chemicals and inoculum are located. The bioreactor is designed to work at very high temperatures (150-1800C), high pressure (377-412 kPa) and also to withstand vacuum which prevents its collapse while cooling.

Types of Bioreactors

Depending on the design of the reactor, the bioreactors are of following types:

Continuous stirred tank bioreactors - These bioreactors have a cylindrical vessel with motor driven central shaft which gives support to one or more agitators (impellers). The shaft is fitted at the bottom of the bioreactor. The diameter of the impeller is usually one third of the vessel diameter. The impellers are available in different designs like- Rustom disc, concave bladded, marine propeller etc. In stirred tank reactors, the air is added to the culture medium under pressure through a device called sparger. The sparger along with the impellers (agitators) enables better and efficient gas distribution through out in the vessel. The advantages of using stirred tank reactors are: the efficient transfer of gas to growing cells which keeps the growth of cells in healthy limits, stirring ensures good mixing of the contents, the operating conditions are flexible and the bioreactors are easily available which makes them commercially viable products.

Bubble column bioreactors - In these bioreactors, the gas or air is introduced at the base of the column through perforated pipes or plates, or metal microporous spargers. The vessel used for bubble column bioreactors is usually cylindrical with an aspect ratio (height o diameter ratio) of 4-6. The rate of flow of gas affects the O2 transfer and mixing.

Airlift bioreactors - Airlift bioreactors are commonly used for aerobic bioprocessing technology. In the airlift bioreactors, the medium of the vessel is divided into two interconnected zones by means of baffle or draft tube. The air/gas is pumped into one of the two zones referred to as 'riser' and the other zone that receives no gas is known as 'downcomer'. The dispersion flows up the riser zone while the down flow occurs in the downcomer. Further there are two types of bioreactors: 1) **Internal loop bioreactor** - These bioreactors have a single container with a central draft tube that creates interior liquid circulation channels which keeps the volume and circulation at a fixed rate for fermentation. (2) **External loop airlift bioreactor**-These have an external loop to keep the liquid in circulation through separate independent channels. The modifications can be made in these bioreactors depending on the requirements of different fermentation processes.(3) **Two stage airlift bioreactors** -These bioreactors have two bioreactors which are basically used for the temperature dependent formation of products. The growing cells from one bioreactor (maintained at temperature 300C are pumped into another bioreactor (at temperature 420C). This is done because it is very difficult to increase the temperature quickly from 300C to 420C in the same vessel. The cells are grown in the first bioreactor and with the help of the fitted valves and a transfer tube and pump, they are transferred into the second bioreactor, where the actual bioprocessing takes place. (4) **Tower bioreactors** - In this type of bioreactor, a high hydrostatic pressure is generated at the bottom of the reactor which increases the solubility of O2 in the medium. Since the top is expanded, the pressure is reduced which helps in the expulsion of CO2. The cycle completes with the medium flowing back into the downcomer. The advantage with Tower bioreactor is that it has high aeration capacities without having moving parts.

Fluidized bed bioreactors - These bioreactors are mainly suitable to carry out reactions involving fluid suspended biocatalysts such as immobilized enzymes, immobilized cells, microbial flocs etc. The design of the bioreactors is such that the top is extended and the reaction column is narrow which retains the solids in the reactor and allows the liquids to flow out. To maintain an efficient operation of fluidized beds, gas is sparged to create a suitable gas-liquid-solid fluid bed. The recycling of the liquid ensures continuous contact between the reaction contents and biocatalysts which increases the efficiency of bioprocessing.

Packed bed bioreactors- A packed bed bioreactor consists of a bed of solid particles, with biocatalysts on or within the matrix of solids, packed in a column. The solids are generally porous or non-porous gels which maybe compressible or rigid in nature. The nutrient broth continuously flows over the immobilized biocatalyst and the products are released into the fluid from where they are removed. However, due to poor mixing, it is difficult to control the pH of packed bioreactors by the addition of acid or alkali.

Photobioreactors - These bioreactors are specialized for fermentation that can be carried out either by exposing to sunlight or artificial illumination. The photobioreactors are made up of glass or transparent plastic which are the solar receivers. The cell cultures are circulated through the solar receivers by using centrifugal pumps or airlift pumps. These bioreactors work in the temperature which ranges from 25-400C. In these bioreactors, the microorganisms e.g. microalgae, cyanobacteria etc. grow during the day time while the products (e.g. beta-carotene, asthaxanthin) are produced during the night.

SOLID SUBSTRATE FERMENTATION (SSF)

In some biotechnological processes, the growth of the microorganisms is carried out on solid substrates more or less in the absence of free water. Only approximately 15% of moisture is present which is essential for solid-substrate fermentation. Cereal grains, wheat bran, sawdust, wood shavings etc are some of the commonly used solid substrates for SSF. The technique of SSF is used for the production of edible mushrooms, cheese, soy sauce, and many enzymes and organic acids. It is carried out as non-aspetic process and therefore saves sterilization costs. The bioreactors used in this type of fermentation process have simple designs with simple aeration process and effluent treatment. The yield of the products is very high, at low energy expenditure. However, in this process only microorganisms, that can tolerate only low moisture content can be used. It also difficult to monitor O2 and CO2 levels in SSF. The slow growth of microorganisms, also become a limiting factor for product formation.

Working of a bioreactor

In the operation of a bioreactor, there are a few steps that are followed.

Sterilization - The most important requirement is to maintain aseptic or sterile conditions for aseptic fermentation. In order to achieve this, the air supplied during fermentation, he growth medium and the bioreactor it self and all it's accessories are sterilized. There are two methods of sterilization that are followed: (1) In situ sterilization- In this, the bioreactor is filled with the required medium followed by injection of pressurized steam into the jacket or coil surrounding the reaction vessel. The whole system is heated to about 1200C and maintained at this temperature for about 20 minutes. However, this method is not energy efficient and prolonged heating destroys the vitamins and precipitates the components of the medium. (2) **Continuous heat sterilization** - In this, the empty bioreactor is first sterilized by injecting pressurized steam and the medium is rapidly heated to 1400C for a short period again by injecting the pressurized steam. This is an energy efficient method and also does not precipitate the medium components.

Aeration - Oxygen is stored in compressed tanks and is introduced at the bottom of the bioreactor through a 'sparger'. Aeration of the fermentation medium supplies oxygen to the production microorganisms and remove carbon dioxide from the bioreactor. The gases released during the fermentation accumulate in the headspace and then pass out through an air outlet. The headspace is a vacant space on the upper part of the bioreactor and is generally about 20% of it's volume. The air-lift system of aeration involves sparging of air done at the bottom of the fermenter with an upward flow of air bubbles. The aeration capacity of the system depends on the air-flow rate and the internal pressure. The stirred system of aeration involves increasing the aeration capacity by stirring using impellers driven by motor. The aeration capacity of the fermenter depends on the rate of stirring, rate of air flow and internal pressure.

Inoculation and sampling - The sterilized bioreactors with growth medium are inoculated with the production organisms. The size of the inoculum is generally 1-10% of the total volume of the medium. During the fermentation process, the samples are regularly withdrawn to check contamination and to measure the amount and quantity of product formed.

Control systems - Various factors like- pH, temperature, dissolved oxygen, adequate mixing, concentration of the nutrients, foam formation etc are continuously monitored to maintain optimal growth environment in the bioreactor. Very sensitive sensors are available which carry out automated monitoring of these variables. The ideal pH range for optimal growth of microorganisms is between 5.5- 8.5. The pH changes due to the release of metabolites into the medium by the growing microorganisms. The required pH level is maintained by adding acid or alkali followed by thorough mixing of the medium components. The optimal temperature is maintained by using the heating and cooling systems fitted in the bioreactor. Continuous monitoring of dissolved oxygen concentration is also a must for the optimal bioreactions. The oxygen is sparingly soluble in water (0.0084g/l at 250C) and is introduced into the bioreactor as the air bubbles.

The concentration of the nutrients is also important because the limiting concentrations of nutrients helps in the optimal product formation and the high nutrient concentrations have inhibitory effect on the microbial growth. Another important factor to control is the "foam formation". The protein rich media is used in industrial fermentation which leads to 'froth' or 'foam formation' on agitation during aeration. Some antifoam chemicals lower the surface tension of the medium and causes the foam bubbles to collapse. Mineral oils with silicone or vegetable oils are also used as antifoam agents. The bioreactors can also be fitted with mechanical foam control devices which break the foam bubbles and throw back into the fermentation medium. The proper and continuous mixing of the microbial culture is very important to maintain optimal levels of oxygen in the nutrient medium and to prevent the accumulation of toxic metabolic products.

Cleaning - After the completion of the fermentation process, the products are 'harvested' (removal of contents for processing) and the bioreactor is prepared for the next round of fermentation after cleaning technically referred to as 'turn around'. The cleaning of the bioreactors is carried out by using high-pressure water jets from the nozzles fitted into the reaction vessel. In order to maintain the cost effectiveness of the bioreactor, the time taken for turn around which is known as 'down time', is kept as short as possible.

DOWNSTREAM PROCESSING (DSP)

The extraction and purification of a biotechnological product from fermentation is referred to as downstream processing. The methods adopted for downstream processing depends on the nature of the end products, it's concentration, stability, and the degree of purification required. Both intracellular metabolites as well as extracellular metabolites are isolated by DSP. The intracellular metabolites are products located within the cells e.g. vitamins, enzymes etc. The **extracellular metabolites** are the products present outside the cells e.g. most antibiotics, amino acids, alcohol, citric acid, enzymes like amylases, proteases etc. some products like vitamin B12, flavomycin etc are present both as intracellular as well as extracellular products.

Downstream processing involves a number of steps which are as follows:

Solid-liquid separation - The first step is to separate whole cells and other insoluble substances from the culture broth. This is done by using several methods:

(1) Flotation- The process of aeration involves the bubbling of gas in to the liquid broth. The cells and other solid particles get adsorbed on gas bubbles which form a foamy layer which is collected and removed.

(2) Flocculation - The cells or cellular debris form large aggregates and settle down which can be easily removed. Some flocculating agents like inorganic salt, organic poly-electrolyte, mineral hydrocolloid etc are often used to achieve appropriate flocculation.

(3) Filtration - this is the most commonly used technique to separate the biomass and culture filtrate. The rate of filtration depends on many factors such as the size of the organism, presence of other organisms, viscosity of the medium, and temperature. Several filters like depth filters, absolute filters, rotary drum vacuum filters, membrane filters etc. are used. There are three major types of filtration processes used depending on the size of the particles- microfiltration, ultrafiltration, and reverse osmosis.

(4) Centrifugation - The centrifugation is mostly used for separating solid particles from liquid phase. The technique of centrifugation is based on the principle of density differences between the particles to be separated and the medium. In the bioreactors, the continuous flow industrial centrifuges are used where there is a continuous feeding of the slurry and collection of clarified fluid. The solid deposits are intermittently removed. Various models of centrifuges used are: Tubular bowl centrifuge, Disc centrifuge, Multichamber centrifuge, Scroll centrifuge or decanter etc.

Release of intracellular products - The biotechnological products that are located with in the cells like vitamins, enzymes, etc are released in an active form for their further processing and isolation. The microorganisms or cells can de disintegrated or disrupted by physical, chemical, or enzymatic methods depending on the nature of the cells.

(1) Physical methods of cell disruption are (i) Ultrasonication, (ii) Osmotic shock(used for releasing hydrolytic enzymes and binding proteins from gram-negative bacteria), (iii) Heat Shock treatment, (iv) High pressure homogenization, (v)Impingement which involves hitting a stationary surface or a second stream of suspended particles with a stream of suspended cells at high velocity and pressure. Microfluidizer is a device developed on the basis of the principle of impingement. (vi) Grinding with glass beads where the cells mixed with glass beads are subjected to a very high speed in a reaction vessel. The cells break as they are forced against the wall of the vessel by the beads

(2) Chemical methods- Treatment with alkalies, organic solvents, and detergents lyse the cells to release the content. Alkali treatment is used for the extraction of some bacterial proteins. The organic solvents like methanol, ethanol, isopropanol, butanol etc also disrupt the cells. The organic solvent, toluene, which is commonly used, dissolves membrane phospholipids and creates the membrane pores to release intracellular contents. The ionic detergents denature the membrane proteins and lyse the cells e.g. cationic-cetyl trimethyl ammonium bromide or anionic-sodium lauryl sulfate. Non-ionic detergents are less reactive and also affect the purification steps.

(3) Enzymatic methods - Lysozyme is the most commonly used enzyme which hydrolyses beta-1,4-glycosidic bonds of the mucopepide in bacterial cell walls e.g. gram positive bacteria. This enzyme is commercially available produced from hen egg white. For gram-negative bacteria, lysozymes are in use with EDTA to break the cells. When the cell wall gets digested by lysozyme, the periplasmic membrane breaks due to osmotic pressure, which releases the intracellular contents. Glucanase and mannanase along with proteases lyse the cell wall of yeast.

Concentration - The biological products are concentrated by getting rid of water which is present in the filtrate. Depending on the nature of the desired products and the cost effectiveness, the techniques used to concentrate the biotechnological products are evaporation, liquid-liquid extraction, membrane filtration, precipitation, and adsorption.

FERMENTERS AND BIOREACTORS DESIGN

There are many requirements that need to be met in the design of a large production scale fermentation facility. Aspects of design to be considered include design yield basis, operating schedule, media sterilization, fermenter and ancillary vessel design, piping systems, CIP/SIP and CGMP compliance. To be successful, a well thought-out and well-designed sanitary/sterile envelope is therefore crucial to the fermentation/biotech facility.

A fermenter is the set up to carry out the process of fermentation. The fermenters vary from laboratory experimental models of one or two litres capacity, to industrial models of several hundred litres capacity, which refers to the volume of the main fermenting vessel.

A **bioreactor** differs from a fermenter in that the former is used for the mass culture of plant or animal cells, instead of micro-organisms. The chemical compounds synthesised by these cultured cells, such as therapeutic agents, can be extracted easily from the cell biomass.

The design engineering and operational parameters of both fermenters and bioreactors are identical. With the involvement of micro-organisms as elicitors in some situations, the distinction between the two concepts is being gradually obliterated.

Functions/Requirements of Fermenter

1. Provide operation free from contamination;
2. Maintain a specific temperature;
3. Provide adequate mixing and aeration;
4. Control the pH of the culture;
5. Allow monitoring and/or control of dissolved oxygen;
6. Allow feeding of nutrient solutions and reagents;
7. Provide access points for inoculation and sampling;
8. Minimize liquid loss from the vessel;
9. Facilitate the growth of a wide range of organisms.

   (Allman A.R., 1999: Fermentation Microbiology and Biotechnology)

Main parts of a Benchtop fermentor

1. Mechanical: drive motor, heater, pump, gas control, etc;
2. Vessel and accessories;
3. Peripheral equipment such as reagent bottles;

4. Instrumentation and sensors.

Types of Fermenter

1. Activated sludge Fermenter
2. Air Lift Fermenter
3. Bubble cap Fermenter
4. Loop Fermenter
5. Mist Fermenter
6. Packed Bed Fermenter
7. Rotating Drug Fermenter
8. Tower Fermenter
9. Trickling Film Fermenter

11

DEVELOPMENT IN FERMENTATION TECHNOLOGY

Fundamental research in enzymology have remarkably developed during the past thirty years. However, the actual application of enzymes has been rather limited for several technical and economical reasons. Such limiting factors are now being overcome year after year. Immobilized enzyme and cell technology might be a kind of break-through in this field. From the industrial view-point, enzyme technology is now coming into its real developmental stage.

Enzyme technology has recently been finding several directions for its development. These directions may be classified as follows;

1. New industrial catalysts,
2. Tools for food production and processing,
3. Pharmaceutical uses,
4. Analytical and measurement tools,
5. Aids for screening new physiologically active substances, and
6. Aids for creation of new sources of energy and raw materials.

RECENT DEVELOPMENTS IN BIOTECHNOLOGY

Category Examples

1. Medicine - Production of antibiotics, steroids, monoclonal antibodies, vaccines, gene therapy, recombinant DNA technology drugs and improvingdiagnosis by enzymes and enzyme sensors.
2. Agriculture - Plant tissue culture, protoplast fusion, introduction of foreign genesinto plants and nitrogen fixation.
3. Chemicals - Organic acids (citric, gluconic) and mineral extraction.
4. Environment- Improvement of waste treatment, replacement of chemicalinsecticides by biological ones and biodegradation of xenobiotics.
5. Food - Single cell protein (SCP), use of enzymes in food processing andfood preservation.
6. Industry - Use of enzymes in detergent industry, textile and energy production

Future Prospects for the Development of Biotechnology

Biotechnology is expected to make an important contribution to the quality of life through a wide range of goods and services in many areas for example:

- Consumer chemicals: adhesives, detergents, dyes, fibers, flavors,gelling and thickening agents, gums, perfumes, pigments, plastics,........ etc.
- Health care (diagnosis and treatment)
- Analytical chemical tools
- Feed stocks (agriculture and animal).
- Energy sources.
- Environment control (air, water and soil)
- Mineral extraction (soil and sea).

CREATIVE FERMENTATION TECHNOLOGY FOR THE FUTURE

Fermentation technology is one of the oldest food technology applications that has been developed and utilized for survival. The origin of Asiaticfermentation technology evolved as early as the littoral foragers period of the Primitive Pottery Age (8,000-3,000 B.C.), which led the Neolithic culture of agriculture in the Northeast Asia. The importance of Primitive Pottery Age for the dietary culture of the region was discussed during the 11thIUFoST World Congress in Seoul (Lee, 2001a). Fermentation technology has developed indigenously all around the world by using natural products from the respective region to produce required food materials, from which the characteristic taste and aroma of each cultural society have been made. Steinkraus(1993) classified fermentation technology of the world which is responsible for manâs survival into six groups: alcoholic, lactic acid, leavened breads, meat substitutes, meat flavored sauces/pastes and protein/flavoring agents fermentation.

The meat eating habit of Western culture needed food preservation technology in order to keep their perishable meat and milk edible for longer period. Meat sausage, cheese and acid fermented milk making were important food preservation technology methods until the time when refrigerators were available in homes. On the other hand, people who ate cereals as staple food in the East wanted to have meaty flavored and salty condiments, which make the bland taste cereal foods more palatable. This demand led Asian people to develop soybean sauce and fish sauce fermentation technology.

Consumer perception of food has changed through history. Until the 20thcentury, the food situation of the world was always problematic. Food shortage could occur any time even in the affluent societies. Therefore, man mostly lived for food, and we may call this earlier period the âsurvival food age.â In the 20th century access to food no longer was a problem in most of the affluent societies, but time was short for work and leisure. People wanted to save time cooking and obtaining food.

Consequentlyconvenience food became the major item in the food market. Now, we all notice that the 21st century is the âfunctional food era.â Health oriented food and nutraceuticals are major concerns of todayâs consumer.

Fermentation technology has adapted itself to social demands. During the survival food age, fermentation was used mainly for food preservation and condiments production. In the convenience food age, it was used for flavorproduction and other ingredients for industrial mass production of food. The 21st century is called the era of tailor made goods satisfying personal demands, and together with the health benefit demands, fermentation technology finds new challenges in the market. Many of the traditional fermented foods are receiving new attention for their health promoting or disease preventing/curing effects. Scientific evidences for their physiological functions are accumulating and the technologies enhancing the beneficial effects are developing rapidly by using modern biotechnological and genetic engineering techniques. Some of the recent developments and future prospects are discussed below.

Lactic acid Fermentation for Probiotics

Yogurt and sour milk products are well known in Western society as probiotic food. Similar products made with rather inexpensive raw materials arefound in many indigenous fermented foods around the world. Korean Kimchi is one of the examples. It is made from vegetables lightly salted (ca. 3%) and fermented by lactic acid bacteria, mainly Leuconostoc mesenteroides. Thelactic acid bacteria grown in kimchi could survive in gastric acid and bile juice reaching to the large intestine (Lee 1997). More important is that kimchi contains various functional components either originated from the ingredients or formed during fermentation as shown in Table 1. The same beneficial effects can be expected from African cereal gruels like ogi and uji, Nigerian gari, and Asian vegetable foods like dhamuoi in Vietnam, dakguadong in Thailand and burong mustala in the Philippines and acid fermented seafoods mixed with cereals (Lee 1994). In addition to the probiotic effect, these vegetable products have excellent prebiotic function.

Combining Western milk fermentation technology and Eastern cereal processing skills led to the creation of fusion foods like soybean yogurt and rice yogurt. Risogurt is an example of lactic acid fermented vegetable drink studied in Korea made from the mixture of rice and soybean protein (Mok,1994). Many of the vegetable yogurt products are found in markets of USA and Japan.

The antimicrobial activity of some lactic acid bacteria against Helicobacter pylori, which is known to cause stomach cancer, is well applied to acid fermented milk beverages and highly accepted by the consumer in Korea (Park et. al. 2001). Anti-diarrhea effect of lactic acid bacteria like Lactobacillus acidophilus is enhanced by protein coating (Chung et al. 2001). Thefreeze dried bacterial cells coated with protein are added to food for preventing diarrhea and other health benefits, and this market is increasing rapidly.

Soybean Fermentation for Cancer and Degenerative disease Prevention

The physiological function and health benefit of soybean is widely recognized today. The blood cholesterol level reduction activity of soybean protein and its hydrolysate and the pseudoestrogenic effect of soybean isoflavones are experimentally proved. In addition to flavor formation, soybean fermentation increases the digestibility and the nutritional value, and also produces health benefit functional compounds (Lee 2001b).

Soybean fermentation for soybean sauce and paste, Korean chongkukjang and Japanese natto production involve the process of enzymatic hydrolysis of protein to make peptides and amino acids. Some peptides in soybean sauce and paste are known to have ACE inhibition, antithrombotic and anti-cancer effects(Shin et. al. 1995, Shon et. al. 1996, Kim 1995). Fermentation with molds, Aspergillus and Rhizopus, converts glycoside form of isoflavones, daidzin and genistin, to aglycones, which have higher potency of estrogenic effect. Production of mucous polysaccharides and enzyme kinase during bacterial (Bacillus) fermentation of soybean is related to the fibrinolytic and immune modulating activity of chongkukjang and natto (Lee et.al. 1991, Sumi et.al. 1987). By enforcing these components through molecular breeding of the microorganisms involved and novel fermentation skills and improved downstream processes, numerous health oriented food products are produced and also under trial in East Asia and will be launched to the world market in nearfuture.

The advantage of mixed culture fermentation

Industrial yogurt production in the West and soybean sauce production in the East were achieved by purely isolated starter culture techniques. However, many of the traditional fermentation starters in the East are naturally fermented mixed culture system. The consumer in their taste and aroma often prefers the traditional fermented products made by the naturally fermented starter culture. The deep and bountiful aroma of Korean rice wine, chongju, and the sharp, strong flavor of Korean soybean sauce, kanjang, are more preferred by Korean people than those of industrial counter parts. The anti-cancer activity of doenjang made from Korean traditional mixed culture starter meju was reported to be higher than that of the industrial product made from koji. The physiological characteristics of microbial strains isolated from traditional fermentation starters and their interactions for growth and biosynthesis of functional compounds are under investigation. More studies on the interaction of different species of microorganism in a fermentation system are needed. Computer aided analysis of microbial interaction will make this sort of study possible and manageable.

Safety issue of fermented products

Since fermentation makes raw food materials edible without cooking, therisks of hazardous microbial contamination always exist in fermented food especially naturally fermented traditional foods. The uneven distribution of salt in lactic acid fermented fish products and contamination of Aspergillus flavus in traditional starter cultures for rice wine and soybean sauce may result in severe food poisoning incidences (Lee 1989, Lee and Lee 2002).

On the other hand, most of the traditional fermentation methods have their own inbuilt safeguard mechanisms. The food infectious and poisoning microorganisms contaminated on the vegetables and other raw materials of kimchi are killed within one week of fermentation period mainly due to the acid formation and bacteriocin production (Lee 1997). Also, large amounts of nitrate and secondary amines in vegetable products are reduced by fermentation(Lee 1986). The importance of fermentation technology for the improvement of hygienic situation of needed region, where cold-chain system is not wellestablished, was discussed at the FAO/WHO Workshop on the Assessment of Fermentation as a Household Technology for Improving Food Safety held in December 11-15, 1995, Pretoria, Republic of South Africa (WHO 1996).

Fermentation technology has confronted new challenge in the era of functional food with its efficient biosynthesis potential. Research for useful strains from traditional fermented foods continues worldwide and relevant information is accumulating. Exchange of knowledge and skills of fermentation technology of the West and the East will accelerate the technology innovation and new product development. Considering the importance of fermentation technology in the era of functional food, the public sectors and international bodies should pay attention to the relevant R&D efforts of academic institutions and industrial research groups. The R&D support of national andinternational funding organizations should include the following research priorities.

1. Nationwide search for useful microbial strains from indigenous fermented products and characterization of their physiological properties.
2. Establishment of a global network on fermentation technology and microbial strains and starter culture information.
3. Studies on the mixed culture fermentation and the interaction of microorganisms in fermentation system.
4. Evaluation of physiological function and health benefit mechanism of the metabolites obtained during fermentation process.
5. Evaluation of safety of fermented products, the microorganisms involved and the metabolites.
6. Evaluation and dissemination of fermentation technology for the improvement of hygienic condition of developing countries.

ADVANCES IN FERMENTATION TECHNOLOGIES

An Industry Overview

This research service on Advances in Fermentation Techniques provides a comprehensive analysis of the technological advancements, emerging trends, drivers, and challenges that affect the industries dependant on fermentation technologies. In addition, this research service provides a technology snapshot of fermentation, an assessment of research and development (R&D) activities by industry (both corporate and institutional), as well as strategic insights of the industry.

This Frost & Sullivan research service titled Advances in Fermentation Technologies—An Industry Overview analyzes and assesses the diverse factors influencing the adoption of fermentation technology across various verticals. It also offers a technology snapshot of the industry and current market applications, assesses R&D activities by corporate and institutional participants, as well as analyzes the technical and business challenges facing the industry. In this research, Frost & Sullivan analysts thoroughly examine the following application sectors: energy and utilities, specialty chemicals, pharmaceuticals, and food and beverage (F&B).

This analysis is available through our Technical Insights program. With this program, clients receive industry-leading market research along with comprehensive, objective information that allows your company to mitigate risk, identify new opportunities, and drive effective strategies for growth.

Technology Overview

Fermentation technology has been a long-favored biological process due to its simplicity of reaction, high specificity, low costs, and versatility of application. Modern industries have complemented the basic principle of fermentation with advances in techniques and genetic engineering to extend its application to the production of a range of biochemicals, biomolecules, and biofuels. Fermentation's applications are rapidly breaching the confines of food stuff such as wine, cheese, bread, and beer, into high-value chemicals, pharmaceutical ingredients, and food-related chemicals. Further, rising petrol prices and depleting fossil fuel reserves have made a robust case for low-cost and simple fermentation processes for producing drop-in chemical replacements. "There has been a spurt in research efforts, both in the corporate and academic sector, driven largely by fermentation applications within the energy and utilities, specialty chemicals, and pharmaceuticals industries," says the analyst of this research. "Energy - or more specifically, biofuels - has evinced considerable interest in applying fermentative techniques to produce cost-competitive bioethanol."

While the uses of fermentation techniques are many in modern industrial processes, there are several challenges to overcome before implementing them in large-scale commercial applications. One of the stand-out challenges is that fermentation is a slow, natural bioprocess, which means that time-scale of operations is often quite long. Companies also have to contend with the low yields and poor process economics in large-scale production. "Although fermentation operating costs are lower than full-scale chemical operations, the costs of raw materials and fermentable sugars remain a challenge for most companies," notes the analyst. "In addition, specially designed enzymes and microbes with specific biosynthetic pathways are also expensive." Therefore, the end products derived will be hard pressed to be cost competitive with other petrochemical-derived alternatives.

To stave off these challenges, companies are increasingly developing robust proprietary microorganisms that can ferment the substrates to produce the desired products. Consequently, companies that are able to provide a solution that combines excellence in microbe technologies with superiority in process technologies will be better equipped to explore new application avenues. Importantly, companies that can take advantage of fermentation's unique bio-capabilities, while still remaining economically feasible, will find themselves operating in a market with expansive possibilities. "Many companies have begun engineering innovative enzymes, microbes, and bacterial strains that can help produce biomolecules in bulk," observes the analyst. "Apart from genetic techniques, some of the recent developments that continue to be explored include process-driven technologies such as solid-state fermentation, consolidated bioprocessing, ultrasonication, syngas fermentation, and dark fermentation." These developments are indicative of the huge potential of fermentation as a low-cost and flexible method to produce a range of end products.

12

Microbial Biomass and Products of Fermentation Process

The soil microbial biomass acts as the transformation agent of the organic matter in soil. As such, the biomass is both a source and sink of the nutrients C, N, P and S contained in the organic matter. It is the center of the majority of biological activity in soil. To properly understand biological activity in soil one must therefore, have knowledge of the microbial biomass. Investigating the flow of C and N in the soil, from newly deposited plant or other materials to the mineral forms of carbon dioxide and ammonium or nitrate ions, clearly shows the central role of the microbial biomass.

The definition of the soil microbial biomass is the living portion of the soil organic matter, excluding plant roots and soil animals larger than 5 x 10 -3 um 3. The microbial biomass generally comprises approximately 2% of the total organic matter in soil and it may be easily dismissed as of minor importance in the soil. However, this chapter will introduce the biomass as an important agent in controlling the overall biological activity of the soil.

Methods of Measuring Microbial Biomass Pool Sizes. Microbial biomass has been measured in soil by a variety of methods. The ideal method would be rapid, sensitive, and distinguish living and active microbial cells from the microbial biomass that is essentially dormant or unresponsi veto the addition of a readily available C source to the soil system. The classical procedure for theestimation of microbial biomass is by direct microscopy. Bio volume of microbial cells in amicroscopic field is determined and the bio volume is then converted to biomass. For conversion of biovolume data to biomass C values, it is necessary to know cell density, cell matter content, and the C content of the dry matter. A conversion factor of 0.13 has been determined assuming the wet density of soil microorganisms is 1.1 g cm-3, a dry matter content of 0.28 g g-1, and a Ccontent of microbial cells of 0.47 g g-1. Using these assumptions we obtain

(1.1 g cm-3)(0.25 g g-1) 0.47 g g-1) = 0. 13 g cm-3

and

biovolume (in cm 3g-1) x 0.13 g cm-3= biomass C (g g-1).

A second procedure is to extract a specific component of the microbial biomass and use the concentration of this component to estimate the concentration of the total biomass. For a specific biomass component to accurately reflect the quantity of the microbial biomass in soil it must meet the following requirements.

1. It must be present in all of the biomass at a known concentration.
2. It must be present in living organisms only and must not accumulate in the soil in nonliving material.
3. It must be extracted quantitatively from soil.
4. It must have a reliable and accurate method of analysis once it is extracted.

Several components have been proposed to estimate and evaluate the microbial biomass in soil.

Probably the most commonly used classical technique for determining the microbial biomass size is that of chloroform fumigation. Fumigation ruptures microbial cells and releases cell walls and cellular contents into the soil. If incubation follows the fumigation, a "flush of decomposition" of the soil organic matter occurs that is due to the decomposition of microorganisms killed during fumigation. When the fumigation and subsequent incubation procedures are conducted under a strict set of conditions, the size of the biomass pool of C can be determined by the size of the carbon dioxide flush using the equation

$$B = F/kc$$

where B is the concentration of biomass C in the soil, F is the amount of carbon dioxide evolved from the fumigated soil minus that from the unfumigated soil, and kc is the fraction of killed biomass C which subsequently is evolved as carbon dioxide. Five assumptions are implicit in the calculation of biomass C from the carbon dioxide flush associated with chloroform fumigation.

1. The C in killed microorganisms is mineralized to carbon dioxide more rapidly than that in living microorganisms.
2. The kill caused by the fumigation is essentially complete.
3. The fraction of microorganisms that die in the unfumigated control is negligible compared to those killed in the fumigated sample.
4. The fraction of the killed biomass that is mineralized (kc) is the same in different soils, i.e. a single kc value can be used to estimate biomass C in a wide variety of soils.
5. Fumigation has no effect on the soil other than the killing of the microbial biomass.

Both chloroform carbon dioxide and methyl bromide have been used as fumigants. Generally with chloroform, the kill is 99% or greater but enough microorganisms survive for an inoculums not to be required. However, an inoculum is required when methyl bromide is used.

A variation of the fumigation method is to do an extraction of the soil after chloroform fumigation. The C content in the extractant is then compared to the amount of C in a similar soil that has not been fumigated. The difference is due to the C released from the microbial biomass. An extraction efficiency (Kec factor) of

0.45 (or a close similar value) is often used to calculate the microbial biomass C value using the fumigation-extraction method.

The chloroform fumigation technique and the application of equations similar to that used for biomass C determinations have also been used to determine the pool sizes of microbial biomass N, P, and S. Microbial S and P measurements have the advantage in that the extraction of the mineral forms of S and P can be carried out immediately after chloroform treatment, eliminating the lengthy delay and unknown mineralization and immobilization rates that occur during incubation. Extraction of C and N constituents within the microbial biomass immediately after chloroform fumigation has not proven as successful.

Another method of determining the size of the microbial biomass includes the initial respiratory response method. This method measures the initial maximum respiration rate in a soil upon addition of an easily degraded C source such as glucose. Micro calorimetry determines the microbial biomass by measuring the rate of heat output from soil and relating this heat output to the size of the microbial biomass.

Dynamics of the Microbial Biomass in Soil. When discussing the microbial biomass in soil, two different situations may be explored. In the first, the biomass size is considered to remain relatively constant over time and C plus other nutrients simply flow through the biomass. Carbon and other nutrients enter the biomass as complex organic forms and leave as carbon dioxide or mineralized forms of the elements.

Organic C, N, P, S <———> microbial <———> CO2, mineral N, biomass mineral P, mineral S

The second case involves the addition of a substrate to soil which stimulates the growth of the microbial biomass and the fate of the nutrients in the organic material may result in their release to the mineral form or they may become incorporated in the expanding microbial biomass pool. Each case must be considered separately.

Studies to understand the role of the microbial biomass other than as a source-sink for nutrients involve knowledge of the activity of the biomass. The term "activity" includes the many processes carried out by microbial enzymes. However, not all of the microbial biomass is operating at the same level of activity and sometimes a direct correlation between the size of the microbial biomass and its activity, as measured by some index, cannot be observed. This is because three different fractions of the microbial biomass, which differ in their activity, exist within the soil. One, the active biomass fraction is that which is capable of growth and all metabolic functions. Two, there exists a sustainable portion of the biomass which is non growing but which can degrade readily available compounds and resume growth under favorable conditions. The third fraction exists of dormant spores or other long-term resting structures.

Estimates of the active biomass fraction range from 10 to 40% of the total identifiable biomass.

Application of Microbial Biomass Measurements to Soil Problems. Information concerning the size of the soil microbial biomass has been used to study soil problems relating to

- The degradation, stabilization, and incorporation of plant C and N into the biomass,
- The effects of freezing and thawing,
- The effect of tillage,
- The role of soil sampling, mixing, and grinding,
- The effect of climatic variations,
- Biomass in forest floors, and
- The effect of faunal feeding.

ALCOHOLIC FERMENTATION PROCESS

Transformation of sugars into alcohols by fermentation is a common practice. Fermentation is carried out by a group of living organisms: yeasts or bacteria. Ethyl alcohol, commonly known as alcohol is one of the most important and popular industrial fermented products. It is a liquid fuel, and can be used as an alternative to automobile fuels. The sugar enriched materials like cane sugar, beet sugar, fruit sugar, potato, com, rice or any other crop of high sugar contents can be used as substrates mainly, along with starchy and ligno-cellulosic materials.

A Materials for fermentation

Following are the types of substrates used for alcehot production.

1. Sugary materials

Examples of sugary materials like sugarcane and its biproducts (bagasse, molasses) and sugarbeet, fruit juice, sweet sorghum, sweet-potatoes etc. Sugarcane molasses is largely being used in many countries for alcohol production. The yield of ethanol is directly proportional to the amount of sugar present.

2. Starchy materials

Tapioca, maize, wheat, barley, oat, sorghum, rice and potatoes are the starchy materials that are used in ethanol production. It has been estimated that 11.7 kg of corn starch can be converted into about 7 liters of ethanol.

3. Lignocellulosic materials

The sources of cellulose and lignocellulosic materials are the agricultural wastes and wood. On the basis of technology available today, about 409 litres of ethanol can be produced from one tonne of lignocellulose.

B. Process of Fermentation

Micro-organisms can grow on carbohydrates such as sugars, under anaerobic conditions. During anaerobic growth, sugars or other fermentable carbohydrates are oxidized and ethyl alcohol is produced as a product of the fermentation process. Optimum temperature requirement for alcohol production is 30-38°C and optimum pH ranges from 3 to 8. Proper balance of macro (nitrogen, phosphorus, potassium, sulphur etc.) and micro (iron, zinc, manganese, copper etc.) nutrients are required for the growth of micro-organisms.

When fermentation is over, which normally takes 4-10 days, the resultant product consists of a mixture of alcohols and water. Ethanol (ethyl alcohol) can be recovered by distillation process i.e. vaporization of alcohol-water mixture.

Production of Biomass

Biomass production can be studied under these sub-heads.

1. Energy plantations

Energy plantation is the practice of planting trees, purely for their use as fuel. Terrestrial **biomass** i.e. the wood plants has been used since long time to generate fire for cooking and other purpose. In recent years, to meet the demand for energy, plantation of energy plants has been re-emphasized. Government has started many plans like afforestation, agri-horticultural practices etc.. It has been estimated that at present, only 1/7th of the world's total energy comes from biomass and a large amount of it remains untapped. In view of getting maximum biomass afforestation, forest management systems will have to be developed. These must include social forestry, silvi-culture tree-use systems, drought tolerant, salt tolerant, pollutant resistant plantations and high density energy plantations.

Annual plants in energy plantations should have fast growth, stress resistance, less palatable to cattle and other animals, high calorific value, absence of deleterious volatiles when smoke comes out, high yield of **biomass** and should be disease and pest free.

2. Social forestry

Plantation through social forestry has been much emphasized by the government of India to meet the demand of fuel and fodder in rural areas. This includes planting trees along road sides, canals, railway lines and wastelands in village. Some of the important plants are Acacia nilotica, Albizia lebbek, Cassia fistula, Eucalpytus globulus, Ficus glome rata, Morus alba, Ferminalia arjuna etc.

3. Silvi-culture energy farms

These are known as short rotation forestry. These silviculture farms employ techniques more similar to agriculture than forestry. The chief objective of energy plantation is to produce biomass from the selected trees and shrub species in the shortest possible time (generally 5-10 years) and at the minimum cost, so as to satisfy local energy needs. This would certainly relieve the pressure on the consumption of fossil fuel like kerosene and prevent the destruction of plant cover, which is one of the primary component of life support system. Due to its high yield per unit area, smaller land requirements for biomass output, shorter life span, increased labour efficiency are the advantageous to adopt silviculture.

13

ROLE OF YEAST IN FERMENTATION PROCESS

Yeasts are eukaryotic micro-organisms classified in thekingdom Fungi, with 1,500 species currently described estimated to be only 1% of all fungal species. Most reproduce asexually by mitosis, and many do so via an asymmetric division process called budding. Yeasts are unicellular, although some species with yeast forms may become multicellular through the formation of a string of connected budding cells known as pseudohyphae, or false hyphae, as seen in most molds. Yeast size can vary greatly depending on the species, typically measuring 3–4 μm indiameter, although some yeasts can reach over 40 μm.

By fermentation the yeast species *Saccharomyces cerevisiae* converts carbohydrates to carbon dioxide and alcohols- for thousands of years the carbon dioxide has been used in bakingand the alcohol in alcoholic beverages. It is also extremely important as a model organism in modern cell biology research, and is one of the most thoroughly researched eukaryoticmicroorganisms. Researchers have used it to gather information about the biology of the eukaryotic cell and ultimately human biology. Other species of yeast, such as*Candida albicans*, are opportunistic pathogens and can causeinfections in humans. Yeasts have recently been used to generate electricity in microbial fuel cells, and produce ethanol for the biofuel industry.

Yeasts do not form a single taxonomic or phylogenet-icgrouping. The term "yeast" is often taken as a synonym for*Saccharomyces cerevisiae*, but the phylogenetic diversity of yeasts is shown by their placement in two separate phyla, theAscomycota and the Basidiomycota. The budding yeasts ("true yeasts") are classified in the order Saccharomycetales.

NUTRITION AND GROWTH

Yeasts are chemoorganotrophs, as they use organic compounds as a source of energy and do not require sunlight to grow. Carbon is obtained mostly from hexose sugars, such as glucose andfructose, or disaccharides such as sucrose and maltose. Some species can metabolize pentosesugars like ribose, alcohols, and organic acids. Yeast species either require oxygen for aerobiccellular respiration (obligate aerobes), or are anaerobic, but also have aerobic methods of energy production (facultative anaerobes). Unlike bacteria, there are no known yeast species that grow only anaerobically (obligate anaerobes). Yeasts grow best in a neutral or slightly acidic pH environment.

Yeasts vary in what temperature range they grow best. For example, *Leucosporidium frigidum* grows at -2 to 20 °C (28 to 68 °F), *Saccharomyces telluris* at 5 to 35 °C (41 to 95 °F) and *Candida slooffi* at 28 to 45 °C (82 to 113 °F). The cells can survive freezing under certain conditions, with viability decreasing over time.

Yeasts are generally grown in the laboratory on solid growth media or in liquid broths. Common media used for the cultivation of yeasts include potato dextrose agar (PDA) or potato dextrose broth, Wallerstein Laboratories nutrient (WLN) agar, yeast peptone dextrose agar (YPD), and yeast mould agar or broth (YM). Home brewers who cultivate yeast frequently use dried malt extract (DME) andagar as a solid growth medium. The antibiotic cycloheximide is sometimes added to yeast growth media to inhibit the growth of *Saccharomyces* yeasts and select for wild/indigenous yeast species. This will change the yeast process.

The appearance of a white, thready yeast, commonly known as kahm yeast, is often a byproduct of the lactofermentation (or pickling) of certain vegetables, usually the result of exposure to air. Although harmless, it can give pickled vegetables a bad flavour and so must be removed regularly during fermentation.

History

The word "yeast" comes to us from Old English *gist*, *gyst*, and from the Indo-European root *yes-*, meaning *boil*, *foam*, or *bubble*. Yeast microbes are probably one of the earliest domesticated organisms. People have used yeast for fermentation and baking throughout history. Archaeologists digging in Egyptian ruins found early grinding stones and baking chambers for yeasted bread, as well as drawings of 4,000-year-old bakeries and breweries. In 1680, the Dutch naturalist Anton van Leeuwenhoek first microscopically observed yeast, but at the time did not consider them to be living organisms, but rather globular structures. In 1857, French microbiologist Louis Pasteur proved in the paper "*Mémoire sur la fermentation alcoolique*" that alcoholic fermentation was conducted by living yeasts and not by a chemical catalyst. Pasteur showed that by bubbling oxygen into the yeast broth, cell growth could be increased, but fermentation was inhibited – an observation later called the "Pasteur effect".

By the late 18th century, two yeast strains used in brewing had been identified: *Saccharomyces cerevisiae*, so called top fermenting yeast, and *S. carlsbergensis*, bottom fermenting yeast. *S. cerevisiae* has been sold commercially by the Dutch for bread making since 1780; while around 1800, the

Germans started producing *S. cerevisiae* in the form of cream. In 1825 a method was developed to remove the liquid so the yeast could be prepared as solid blocks. The industrial production of yeast blocks was enhanced by the introduction of the filter press in 1867. In 1872, Baron Max de Springer developed a manufacturing process to create granulated yeast, a technique that was used until the first World War. In the United States, naturally occurring airborne yeasts were used almost exclusively until commercial yeast was marketed at the Centennial Exposition in 1876 in Philadelphia, where Charles L. Fleischmann exhibited the product and a process to use it, as well as serving the resultant baked bread.

Reproduction

The yeast cell's life cycle:

1. Budding
2. Conjugation
3. Spore

Yeasts, like all fungi, may have asexual and sexualreprod-uctive cycles. The most common mode of vegetative growth in yeast is asexual reproduction bybudding. Here a small bud (also known as a bleb), or daughter cell, is formed on the parent cell. Thenucleus of the parent cell splits into a daughter nucleus and migrates into the daughter cell. The bud continues to grow until it separates from the parent cell, forming a new cell. Some yeasts, including*Schizosaccharomyces pombe*, reproduce by fissioninstead of budding.

Under high stress conditions, haploid cells will generally die; under the same conditions, however,diploid cells can undergo sporulation, entering sexual reproduction (meiosis) and producing a variety of haploid spores, which can go on to mate (conjugate), reforming the diploid.

Some pucciniomycete yeasts, particularly species of *Sporidiobolus* and *Sporobolomyces* produce aerially dispersed, asexual ballistoconidia.

Uses

The useful physiological properties of yeast have led to their use in the field of biotechnology.Fermentation of sugars by yeast is the oldest and largest application of this technology. Many types of yeasts are used for making many foods: baker's yeast in bread production; brewer's yeast in beerfermentation; yeast in wine fermentation and for xylitol production. So-called red rice yeast is actually a mold, *Monascus purpureus*. Yeasts include some of the most widely used model organismsfor genetics and cell biology.

Alcoholic Beverages

Alcoholic beverages are defined as beverages that contain ethanol (C2H5OH). This ethanol is almost always produced by fermentation – the metabolism of carbohydrates by certain species of yeast under anaerobic or low-oxygen conditions. Beverages such as wine, beer, or distilled spirits all use yeast at some stage of their production. A distilled beverage is a beverage containing ethanol that has been purified by distillation. Carbohydrate-containing plant material is fermented by yeast, producing a dilute solution of ethanol in the process. Spirits such as whiskey and rum are prepared by distilling these dilute solutions of ethanol. Components other than ethanol are collected in the condensate, including water, esters, and other alcohols, which account for the flavour of the beverage.

Beer

Brewing yeasts may be classed as "top cropping" (or "top fermenting") and "bottom cropping" (or "bottom-fermenting"). Top cropping yeasts are so called because they form a foam at the top of the wort during fermentation. An example of a top cropping yeast is *Saccharomyces cerevisiae*, sometimes called an "ale yeast". Bottom cropping yeasts are typically used to produce lager-type beers, though they can also produce ale-type beers. These yeasts ferment well at low temperatures. An example of bottom cropping yeast is *Saccharomyces pastorianus*, formerly known as *S. carlsbergensis*.

Decades ago, taxonomists reclassified S. carlsbergensis (uvarum) as a member of S. cerevisae, noting that the only distinct difference between the two is metabolic. Lager strains of S. cerevisae secrete an enzyme called melibiase, allowing it to hydrolyse the disaccharide melibiose into more fermentable monosaccharides. Top cropping and bottom cropping, cold fermenting and warm fermenting distinctions are largely generalizations used to by the laymen to communicate to the general public. For more on the taxonomical differences, see Terrance M. Dowhanick, B.SC., PH. D "Yeast - Strains and Handling Techiques" in The Practical Brewer, a publication of the Master Brewers Association of the Americas.

The most common top cropping brewer's yeast, *S. cerevisiae*, is the same species as the common baking yeast. Brewer's yeast is also very rich in essential minerals and the B vitamins (except B12). However, baking and brewing yeasts typically belong to different strains, cultivated to favour different characteristics: baking yeast strains are more aggressive, to carbonate dough in the shortest amount of time possible; brewing yeast strains act slower, but tend to produce fewer off-flavours and tolerate higher alcohol concentrations (with some strains, up to 22%).

Dekkera/Brettanomyces is a genus of yeast known for their important role in the production of *Lambic*and specialty *sour ales*, along with the secondary conditioning of a particular Belgian Trappist beer. The taxonomy of the genus Brettanomyces has been debated since its early discovery and has seen many re-classifications over the years. Early classification was based on a few species that reproduced asexually (anamorph form) through multipolar budding . Shortly after, the formation of ascospores was observed and the genus Dekkera, which reproduces sexually (teleomorph form), was introduced as part of the taxonomy. The current taxonomy includes five species within the genera of Dekkera/Brettanomyces. Those are the anamorphs *Brettanomyces bruxellensis*, *Brettanomyces anomalus*,

Brettanomyces custersianus, *Brettanomyces naardenensis*, and *Brettanomyces nanus*, with teleomorphs existing for the first two species, *Dekkera bruxellensis* and *Dekkera anomala*. The distinction between Dekkera and Brettanomyces is arguable with Oelofse et al. (2008) citing Loureiro and Malfeito-Ferreira from 2006 when they affirmed that current molecular DNA detection techniques have uncovered no variance between the anamorph and teleomorph states. Over the past decade, Brettanomyces spp. have seen an increasing use in the craft-brewing sector of the industry with a handful of breweries having produced beers that were primary fermented with pure cultures of Brettanomyces spp. This has occurred out of experimentation as very little information exists regarding pure culture fermentative capabilities and the aromatic compounds produced by various strains. Dekkera/Brettanomyces spp. have been the subjects of numerous studies conducted over the past century although a majority of the recent research has focused on enhancing the knowledge of the wine industry. Recent research on 8 Brettanomyces strains available in the brewing industry focused on strain specific fermentations and identified the major compounds produced during pure culture anaerobic fermentation in wort.

Wine

Yeast is used in winemaking, where it converts the sugars present in grape juice (must) into ethanol. Yeast is normally already present on grape skins (the white powder called "the bloom"). Fermentation can be done with this endogenous "wild yeast," but this procedure gives unpredictable results, which depend upon the exact types of yeast species present. For this reason, a pure yeast culture is usually added to the must; this yeast quickly dominates the fermentation. The wild yeasts are repressed, which ensures a reliable and predictable fermentation.

Most added wine yeasts are strains of *S. cerevisiae*, though not all strains of the species are suitable. Different *S. cerevisiae* yeast strains have differing physiological and fermentative properties, therefore the actual strain of yeast selected can have a direct impact on the finished wine. Significant research has been undertaken into the development of novel wine yeast strains that produce atypical flavour profiles or increased complexity in wines.

The growth of some yeasts, such as *Zygosaccharomyces* and *Brettanomyces*, in wine can result inwine faults and subsequent spoilage. *Brettanomyces* produces an array of metabolites when growing in wine, some of which are volatile phenolic compounds. Together, these compounds are often referred to as "*Brettanomyces* character", and are often described as "antiseptic" or "barnyard" type aromas. *Brettanomyces* is a significant contributor to wine faults within the wine industry.

Researchers from University of British Columbia, Canada, have found a new strain of yeast that has reduced amines. The amines in red wine and Chardonnay produce off-flavors and cause headaches and hypertension in some people. About 30 percent of people are sensitive to biogenic amines, such as histamines.

Baking

Yeast, most commonly *S. cerevisiae*, is used in baking as a leavening agent, where it converts thefermentable sugars present in dough into the gas carbon dioxide. This causes the dough to expand or rise as gas forms pockets or bubbles. When the dough is baked, the yeast dies and the air pockets "set", giving the baked product a soft and spongy texture. The use of potatoes, water from potato boiling, eggs, or sugar in a bread dough accelerates the growth of yeasts. Most yeasts used in baking are of the same species common in alcoholic fermentation. Additionally, *Saccharomyces exiguus*(also known as *S. minor*), a wild yeast found on plants, fruits, and grains, is occasionally used for baking. Sugar and vinegar provide the best conditions for yeast to ferment. In bread making, the yeast initially respires aerobically, producing carbon dioxide and water. When the oxygen is depleted,anaerobic respiration begins, producing ethanol as a waste product; however, this evaporates during baking.

It is not known when yeast was first used to bake bread. The first records that show this use came from Ancient Egypt. Researchers speculate a mixture of flour meal and water was left longer than usual on a warm day and the yeasts that occur in natural contaminants of the flour caused it to ferment before baking. The resulting bread would have been lighter and tastier than the normal flat, hard cake.

Today, there are several retailers of baker's yeast; one of the best-known in North America is Fleischmann's Yeast, which was developed in 1868. During World War II, Fleischmann's developed a granulated active dry yeast, which did not require refrigeration and had a longer shelf lifethan fresh yeast. The company created yeast that would rise twice as fast, reducing baking time. Baker's yeast is also sold as a fresh yeast compressed into a square "cake". This form perishes quickly, and must therefore be used soon after production. A weak solution of water and sugar can be used to determine if yeast is expired. In the solution, active yeast will foam and bubble as it ferments the sugar into ethanol and carbon dioxide. Some recipes refer to this as proofing the yeast as it "proves" (tests) the viability of the yeast before the other ingredients are added. When using a sourdough starter, flour and water are added instead of sugar; this is referred to as proofing the sponge.

When yeast is used for making bread, it is mixed with flour, salt, and warm water or milk. The dough is kneaded until it is smooth, and then left to rise, sometimes until it has doubled in size. Some bread doughs are knocked back after one rising and left to rise again. A longer rising time gives a better flavour, but the yeast can fail to raise the bread in the final stages if it is left for too long initially. The dough is then shaped into loaves, left to rise until it is the correct size, and then baked. Dried yeast is usually specified for use in a bread machine; however, a (wet) sourdough starter can also work.

YEAST FERMENTATION TECHNOLOGIES

The yeast strain, *Saccharomyces cerevisiae*, is a species of budding yeast commonly used as a leavening in baking and a microorganism for fermentation. The strain of *S. cerevisiae* is known as a model organism for the study of human biology providing valuable information related to cell growth and genetics. Additionally, the microorganism supports industrial applications as the leading fermenting yeast in the biological and medicinal sciences producing commercially available proteins-enzymes and co-enzymes.

A Long History of Yeast Fermentation Technology

Oriental Yeast Company (OYC) was established in 1929 as the first company in Japan to produce yeast for baking. As the need for yeast grew both in Japan and worldwide, OYC grew as well-producing yeast for several applications including baking powders, laboratory animal feeds, and aquacultural feeds. OYC also focused on the establishment of yeast-derived products for medical diagnostic applications including enzymes, coenzymes, antibodies, recombinant proteins, and clinical chemistry reagents.

Committed to quality products and customer satisfaction, OYC has continued to grow to better serve its global market. Strictly maintaining its ISO 9000 accreditation, it has entered into several partnerships and technology ventures worldwide-establishing offices in the United States in 1994 as well as Europe and China in 2008. In addition to yeast for baking and bakery ingredients, OYC has served diagnostic, pharmaceutical, and nutritional industries as well as contract labs, research institutes and universities for over 30 years.

OYC's worldwide network of products-supported by close client relationships and excellent customer service-ensures on time shipments, up-to-date technical information, and complete sales support.

Setting benchmarks for best practices and industry standards has been an important part of OYC's history. OYC's two research ventures focus on both the nutritional and scientific applications of yeast fermentation science:

Tokyo Laboratory of Yeast and Fermentation was founded in tandem with OYC in 1929. Ever since, the laboratory has played a large role in the development and research of baker's yeast in Japan. The Tokyo Laboratory has isolated over 13,000 strains of bakers' yeast and helped bring 30 varieties to the world market.

Nagahama Institute for Biochemical Science was founded in 1987 to establish research and development, putting gene technology and other related technologies to practical use for analysis, testing, and production. With diverse areas of research-antiserum studies based on recombinant gene technology, protein engineering, and gene analysis-the Institute developed a series of raw materials and procedures that are used worldwide

Utilisation of Yeast for Baking

- Bread came into being from observations made by the Egyptians 3,000 years ago of the transformation of the dough by beer froth which they scooped up from the surface of bowls. Since then, the history of bread, yeast and men is tightly linked. For a long time, bread was a basic food element in Europe and in the whole Mediterranean basin, with a high symbolic content.
- Without fermentation by yeast, a living and natural product, we could not obtain the equivalent of bread or Viennese pastry.
- Bread is less important nowadays as a basic food in developed countries, but it is still a healthy product, harmoniously helping to bring about a balanced diet and nutritionists recommend its consumption.

Baking Fermentation

- In the process of fermentation, yeast produces carbon dioxide, alcohol and other compounds which enable dough to rise and modify its physical properties. When the fermentation is correctly achieved, depending on the quality of flour among other things, the baker will obtain the proper external and internal characteristics (grain and texture) suitable for a determined finished product.
- Together with alcohol, a small amount of other volatile compounds are released. These are involved in the original taste and flavours of baked leavened products. Most of these compounds, and all of the alcohol, evaporate when baked.
- There are two stages during the fermentation of a dough made from flour, water, salt and yeast, and with no extra sugar added.
- First of all, yeast ferments sugars naturally present in flour, which can be directly and easily assimilated. These sugars represent about 1.5% of the flour weight. At the end of this first stage, gaseous releases more or less slow down.
- The second stage corresponds to the fermentation of a sugar contained in flour called maltose. Maltose comes from the action of some enzymes, the amylases, on the starch granules of the flour, damaged during the milling process. Amylases which are naturally present in flour, split starch into small fractions of a much simpler sugar, the maltose. The action of amylases starts as soon as water is added to the flour and stops during baking.
- The action of the flour amylases is completed by that of another enzyme of yeast, the maltase which, in its turn, splits maltose to give the most simple sugar, glucose. The glucose is transformed by the yeast into carbon dioxide and alcohol.
- Maltose formed from starch must be present in sufficient quantity so that the production of carbon dioxide makes dough rise correctly until it is put in the oven.
- Since 1960, the bakery industry has benefited from better adapted yeast strains, allowing a rapid process as they ferment the maltose sugar earlier. Effectively, the maltose is fermented only when there are no or little pre-existing sugars or added simple sugars left in the dough.

- The yeast industry has studied to a great extent baker's yeast strains, not only with a view to making them more rapid, but also to adapt them as much as possible to different types of baking processes.
- If sucrose has been incorporated into the dough, it is immediately transformed by a yeast enzyme, the invertase, into glucose and fructose (the two basic links composing sucrose are thus separated).
- When sucrose or glucose are added to the dough, they are directly fermented before maltose. This means that in such a dough, mainly sucrose or glucose are consumed by the yeast. These sugars do not contribute to giving the bread a sweet taste since they are partly consumed.

Use of Yeast

The Dosage of Yeast

For breadmaking, the amount of yeast used is between 2 and 5 kilos of compressed yeast for 100 kilos of flour. In practice, this proportion varies effectively according to the baking process and the temperature of the bakery.

In 100 grams of flour, 1 to 10 million micro-organisms live among which only 30,000 are the so-called wild yeasts. A standard amount of 2.5 grams of baker's yeast for 100g of flour provides 25 billion yeast cells. This is the proof that baker's yeast is predominant in the bakery fermentation.

Fermentation

Fermentation begins as soon as the yeast is in contact with the mixture of water and flour. During the first fermentation (bulk fermentation), the baker lets the dough rise for a first time. During that stage, the physical properties of the dough (extensibility and elasticity) are modified, thus completing the kneading action.

This is an important stage on which depends the final quality of the bread: external and internal characteristics, taste and aroma.

The mechanisation of the division of the dough requires some rest so that the dough pieces may relax before moulding, in order to recover some extensibility.

The fermentation process goes on in the cut lumps of dough during the proofing which can be made on layers, on flexible nets or in moulds, in cupboards at ambient temperature or in proofing ovens. In order to prevent the lumps of dough drying out during this stage, they must be covered so as not to hinder the development and the appearance of the finished product.

This stage can be achieved by a cold-controlled fermentation, in chambers in which temperature and degree of humidity (hygrometric degree) are determined with precision. Thus, the fermentation can be controlled and working hours organised.

In the case of French bread, the baker makes small and regular blade incisions in the lumps of dough just before putting them in the oven. These incisions will give nice edges, the puckering, on the crust of the bread. It is a delicate operation requiring much experience: it is said that it is the baker's signature because it essentially depends on his know-how.

Without these blade incisions, the loaf of bread loses its shape and bursts under the violent release of the carbon dioxide and steam which appear right after being put into the oven. The blade incision is therefore not made for decorative reasons only, but it also gives the loaf a regular aspect by creating a zone where the dough will be torn in the right direction. Thus, the loaf gets a regular volume and an optimal development.

Science and History

Bakers know two ways of getting dough to rise :

The first one, only for biscuits and cakes, is the use of a chemical powder which makes dough expand as soon as it is put into the oven ;

The second one, for bread and Viennese pastry, is the use of natural living yeast which requires fermentation before it is put into the oven in order for the dough to develop.

The use of chemical powder may seem more simple and convenient, but its use is not appropriate for bread making. On the one hand, carbon dioxide is not produced in sufficient quantity to obtain a light and alveolated bread texture and, on the other hand, only a fermentation agent like yeast can give bread its organoleptic quality.

Bread is symbolical food which, in spite of its various compositions, is appreciated worldwide thanks to a special manufacturing process : fermentation.

Between 1857 and 1863 Louis Pasteur demonstrated the role of yeast as a micro-organism responsible for fermentation.

" All yeast which ferment bread, beer, wine and cider correspond to microscopic living cells of a microscopic fungus, Saccharomyces cerevisiae.

Many varieties of Saccharomyces cerevisiae exist in nature and they are more or less suitable for the various types of fermentation".

Yeast's Crucial Roles in Breadbaking

It acts as a leavener, dough developer, and flavor builder

Yeast is the driving force behind fermentation, the magical process that allows a dense mass of dough to become a well-risen loaf of bread. And yet yeast is nothing more than a single-celled fungus. How does it do it?

Yeast works by consuming sugar and excreting carbon dioxide and alcohol as byproducts. In bread making, yeast has three major roles. Most of us are familiar with yeast's leavening ability. But you may not be aware that fermentation helps to strengthen and develop gluten in dough and also contributes to incredible flavors in bread.

Yeast makes dough rise

Yeast cells thrive on simple sugars. As the sugars are metabolized, carbon dioxide and alcohol are released into the bread dough, making it rise. Photo by: Scott Phillips.

The essentials of any bread dough are flour, water, and of course yeast. As soon as these ingredients are stirred together, enzymes in the yeast and the flour cause large starch molecules to break down into simple sugars. The yeast metabolizes these simple sugars and exudes a liquid that releases carbon dioxide and ethyl alcohol into existing air bubbles in the dough.

If the dough has a strong and elastic gluten network, the carbon dioxide is held within the bubble and will begin to inflate it, just like someone blowing up bubblegum. As more and more tiny air cells fill with carbon dioxide, the dough rises and we're on the way to leavened bread.

Yeast Strengthens Bread Dough

When you stir together flour and water, two proteins in the flour—glutenin and gliadin—grab water and each other to form a bubblegum-like, elastic mass of molecules that we call gluten. In bread making, we want to develop as much gluten as we can because it strengthens the dough and holds in gases that will make the bread rise.

Once flour and water are mixed together, any further working of the dough encourages more gluten to form. Manipulating the dough in any way allows more proteins and water to find each other and link together. If you've ever made homemade pasta, you know that each time you roll the dough through the machine, the dough becomes more elastic; in other words, more gluten is developed. And with puff pastry dough, every time you fold, turn, and roll the dough, it becomes more elastic.

Yeast, like kneading, helps develop the gluten network. With every burst of carbon dioxide that the yeast releases into an air bubble, protein and water molecules move about and have another chance to connect and form more gluten. In this way, a dough's rising is an almost molecule-by-molecule kneading. Next time you punch down bread dough after its first rise, notice how smooth and strong the gluten has become, in part from the rise.

At this stage, most bakers stretch and tuck the dough into a round to give it a smooth, tight top that will trap the gases produced by fermentation. Then they let this very springy dough stand for 10 to 15 minutes. This lets the gluten bonds relax a little and makes the final shaping of the dough easier. This rounding and resting step isn't included in many home baking recipes, but it's a good thing to do.

Fermentation generates flavor in bread

As Harold McGee, the author of On Food & Cooking, has pointed out, big molecules in proteins, starches, and fats don't have much flavor, but when they break down into their building blocks—proteins into amino acids, starches into sugars, or fats into free fatty acids—they all have marvelous flavors. Fermentation, whether it's acting on fruit juices to make wine or on flour to make bread, does exactly that—it breaks down large molecules into smaller, flavorful ones.

At the beginning of fermentation, enzymes in the yeast start breaking down starch into more flavorful sugars. The yeast uses these sugars, as well as sugars already present in the dough, and produces not only carbon dioxide and alcohol but also a host of flavorful byproducts such as organic acids and amino acids. A multitude of enzymes encourages all kinds of reactions that break big chains of molecules into smaller ones—amylose and maltose into glucose, proteins into amino acids.

As fermentation proceeds, the dough becomes more acidic. This is due in part to rising levels of carbon dioxide, but there are also more flavorful organic acids like acetic acid (vinegar) and lactic acid being formed from the alcohol in the dough. (This is similar to what happens to a bottle of wine that has been left uncorked for a while: the alcohol combines with oxygen to make vinegar.) The acidity of the dough causes more molecules to break down. The dough becomes a veritable ferment of reactions. Eventually, the amount of alcohol formed starts to inhibit the yeast's activity.

Yeast has help in producing flavorful compounds. Bacteria are important flavor builders as well. There are bacteria in the dough from the beginning, but as long as the yeast is very active, it consumes sugars as quickly as they're produced, leaving no food for the bacteria, which also like sugar. But when bakers chill a dough and slow down its rise, the cold dramatically reduces yeast activity. The bacteria, on the other hand, function well even in cold temperatures, so they now have an opportunity to thrive, producing many more marvelously flavorful acids.

Baker's Yeast Production-Principles

Yeasts can grow in the presence or absence of air. Anaerobic growth, growth in the absence of oxygen, is quite slow and inefficient. For instance, in bread dough, yeast grow very little. Instead, the sugar that can sustain either fermentation or growth is used mainly to produce alcohol and carbon dioxide. Only a small portion of the sugar is used for cell maintenance and growth. In contrast, under aerobic conditions, in the presence of a sufficient quantity of dissolved oxygen, yeast grow by using most of the available sugar for growth and producing only negligible quantities of alcohol.

This means that the baker who is interested in the leavening action of carbon dioxide works under conditions that minimize the presence of dissolved oxygen. On the other hand, a yeast manufacturer that wants to produce more yeast cell mass, works under aerobic conditions by bubbling air through the solution in which the yeast is grown.

The problem posed to the yeast manufacturer, however, is not as simple as just adding air during the fermentation process. If the concentration of sugar in the fermentation growth media is greater than a very small amount, the yeast will produce some alcohol even if the supply of oxygen is adequate or even in abundance. This problem can be solved by adding the sugar solution slowly to the yeast throughout the fermentation process. The rate of addition of the sugar solution must be such that the yeast uses the sugar fast enough so that the sugar concentration at any one time is practically zero. This type of fermentation is referred to as a fed-batch fermentation.

Baker's Yeast Production Practice

The baker's yeast production process flow chart attached below can be divided into four basic steps. In order these steps are, molasses and other raw material preparation, culture or seed yeast preparation, fermentation and harvesting and filtration and packaging. The process outlined in the flow chart takes approximately five days from start to finish.

The basic carbon and energy source for yeast growth are sugars. Starch can not be used because yeast does not contain the appropriate enzymes to hydrolyze this substrate to fermentable sugars. Beet and cane molasses are commonly used as raw material because the sugars present in molasses, a mixture of sucrose, fructose and glucose, are readily fermentable. In addition to sugar, yeast also require certain minerals, vitamins and salts for growth. Some of these can be added to the blend of beet and cane molasses prior to flash sterilization while others are fed separately to the fermentation. Alternatively, a separate nutrient feed tank can be used to mix and deliver some of the necessary vitamins and minerals. Required nitrogen is supplied in the form of ammonia and phosphate is supplied in the form of phosphoric acid. Each of these nutrients is fed separately to the fermentation to permit better pH control of the process. The sterilized molasses, commonly referred to as mash or wort, is stored in a separate stainless steel tank. The mash stored in this tank is then used to feed sugar and other nutrients to the appropriate fermentation vessels.

Baker's yeast production starts with a pure culture tube or frozen vial of the appropriate yeast strain. This yeast serves as the inoculum for the pre-pure culture tank, a small pressure vessel where seed is grown in medium under strict sterile conditions. Following growth, the contents of this vessel are transferred to a larger pure culture fermentor where propagation is carried out with some aeration, again under sterile conditions. These early stages are conducted as set-batch fermentations. In a set-batch fermentation all the growth media and nutrients are introduced to the tank prior to inoculation.

From the pure culture vessel, the grown cells are transferred to a series of progressively larger seed and semi-seed fermentors. These later stages are conducted as fed-batch fermentations. During a fed-batch fermentation, molasses, phosphoric acid, ammonia and minerals are fed to the yeast at a controlled rate. This rate is designed to feed just enough sugar and nutrients to the yeast to maximize multiplication and prevent the production of alcohol. In addition, these fed-batch fermentations are not completely sterile. It is not economical to use pressurized tanks to guarantee sterility of the large volumes of air required in these fermentors or to achieve sterile conditions during all the transfers through the many pipes, pumps and centrifuges. Extensive cleaning of the equipment, steaming of pipes and tanks and filtering of the air is practiced to insure as aseptic conditions as possible.

At the end of the semi-seed fermentation, the contents of the vessel are pumped to a series of separators that separate the yeast from the spent molasses. The yeast is then washed with cold water and pumped to a semi-seed yeast storage tank where the yeast cream is held at 34 degrees Fahrenheit until it is used to inoculate the commercial fermentation tanks. These commercial fermentors are the final step in the fermentation process and are often referred to as the final or trade fermentation.

Commercial fermentations are carried out in large fermentors with working volumes up to 50,000 gallons. To start the commercial fermentation, a volume of water, referred to as set water, is pumped into the fermentor. Next, in a process referred to as pitching, semi-seed yeast from the storage tank is transferred into the fermentor. Following addition of the seed yeast, aeration, cooling and nutrient additions are started to begin the 15-20 hour fermentation. At the start of the fermentation, the liquid seed yeast and additional water may occupy only about one-third to one-half of the fermentor volume. Constant additions of nutrients during the course of fermentation bring the fermentor to its final volume. The rate of nutrient addition increases throughout the fermentation because more nutrients have to be supplied to support growth of the increasing cell population. The number of yeast cells increase about five- to eight-fold during this fermentation.

Air is provided to the fermentor through a series of perforated tubes located at the bottom of the vessel. The rate of airflow is about one volume of air per fermentor volume per minute. A large amount of heat is generated during yeast growth and cooling is accomplished by internal cooling coils or by pumping the fermentation liquid, also known as broth, through an external heat exchanger. The addition of nutrients and regulation of pH, temperature and airflow are carefully monitored and controlled by computer systems during the entire production process. Throughout the fermentation, the temperature is kept at approximately 86 degrees Fahrenheit and the pH in the range of 4.5-5.5.

At the end of fermentation, the fermentor broth is separated by nozzle-type centrifuges, washed with water and re-centrifuged to yield a yeast cream with a solids concentration of approximately 18%. The yeast cream is cooled to about 45 degrees Fahrenheit and stored in a separate, refrigerated stainless steel cream tank. Cream yeast can be loaded directly into tanker trucks and delivered to customers equipped with an appropriate cream yeast handling system. Alternatively, the yeast cream can be pumped to a plate and frame filter press and dewatered to a cake-like consistency with a 30-32% yeast solids content. This press cake yeast is crumbled into pieces and packed into 50-pound bags that are stacked on a pallet. The yeast heats up during the pressing and packaging operations and the bags of crumbled yeast must be cooled in a refrigerator for a period of time with adequate ventilation and placement of pallets to permit free access to the cooling air. Palletized bags of crumbled yeast are then dis tributed to customers in refrigerated trucks.

14

Safety Analysis of Fermented Foods

Fermentation is one of the oldest technologies used for food preservation. Over the centuries, it has evolved and been refined and diversified. Today, a variety of food products is derived from this technology in households, small-scale food industries as well as in large enterprises. Furthermore, fermentation is an affordable food preservation technology and of economic importance to developing countries.

Ethyl carbamate (EC), also known as urethane, is a chemical contaminant naturally formed in fermented foods during the fermentation process or during storage. Measurable levels of EC have been found in foods such as bread, soy sauce and yogurt, and in alcoholic beverages such as spirits, wine and beer. The health concern of EC in food is related to its carcinogenic potential. Ethyl carbamate has a history of use in industry, medicine and veterinary applications. Its use in human medicine was later banned due to toxicological concerns and lack of efficacy.

The levels of EC present in different alcoholic beverages and fermented foods have been extensively reviewed in overseas countries. Studies demonstrated that foods produced at least partially by yeast fermentation seemed to be much more likely to contain EC than those fermented by lactic acid bacteria, acetic acid bacteria and moulds, while the non-yeast fermented products (e.g. cheese) usually contained low to non-detectable levels of EC.

There is currently no international standard for the maximum allowable level of EC in foods. However, some countries have established maximum levels of EC in alcoholic beverages. Canada was the first country to introduce maximum levels for ethyl carbamate in a variety of alcoholic beverages, from 30 ìg/L for wine to 400 ìg/L for fruit brandies. The US has voluntary target levels of EC for her national food products, and has notified all countries exporting wines to the US that they must develop programs to meet these target levels. There are currently no harmonized maximum levels of EC in foods in the European Union, although some member states have recommended maximum EC levels in alcoholic beverages. Recently, Korea has also established a maximum EC level of 30 ìg/L in wine.

At present, there is no subsidiary legislation governing the maximum level of EC in fermented foods and alcoholic beverages in Hong Kong. While previous studies carried out overseas mainly focused on food items of the Western diet, no comprehensive studies on the level of contamination by EC in commonly consumed local fermented foods and beverages have been conducted. The health risk associated with dietary exposure to EC upon consumption of fermented foods and beverages for the local population is not clear.

It is important that a risk assessment study on EC be conducted, particularly on the local fermented foods such as preserved vegetables, fermented soy products, local condiments and sauces, and alcoholic beverages including Chinese fermented wines. The study would be of value to both the local and international community, for providing first-hand information on the level of contamination of EC in the commonly consumed local fermented foods in Hong Kong, for assessment of the associated health risk of dietary exposure to EC of the local population, and for submission of the EC contamination data to the JECFA database.

HAZARD IDENTIFICATION

Ethyl carbamate is the ester of carbamic acid. It is naturally formed in fermented foods mainly as a byproduct of fermentation, primarily from the reaction of alcohol (ethanol) with urea and its break-down products. Cyanate is probably the ultimate precursor in most cases, reacting with ethanol to form the carbamate ester.

Ethyl carbamate can be formed from various substances derived from fermented foods and beverages, including urea, hydrogen cyanide, citrulline and other N-carbamyl compounds. For example, in stone fruit spirits, cyanogenic glycosides from the stones may be degraded through enzymatic action to cyanide which is then oxidised to cyanate and reacts with ethanol to form EC. Another example is the formation of urea from the degradation of arginine by yeast fermentation, which is then broken down to isocyanate and reacts with ethanol to form EC. The chemical reaction between urea and ethanol has been found to accelerate exponentially with increase in temperature. Besides temperature, the presence of light and duration of storage are the other two key factors influencing the formation of ethyl carbamate in fermented foods

The main source of dietary exposure to EC in human population is through the consumption of fermented foods and beverages containing EC. Alcoholic beverage is the known main contributor to EC exposure.

Hazard Characterisation

The Joint Food and Agriculture Organization/World Health Organization Expert Committee on Food Additives (JECFA) evaluated EC in 2005 and concluded that the chemical was genotoxic and a multisite carcinogen in all animal species tested and a potential carcinogen in human.

In 2007, the International Agency for Research on Cancer (IARC) re-classified EC as "probably carcinogenic to humans".

Kinetics and Metabolism

Ethyl carbamate is absorbed rapidly and almost completely from the gastrointestinal tract and is evenly distributed throughout the body. It is also eliminated quickly. In mice, more than 90% of the absorbed EC is eliminated as carbon dioxide within 6 hours

Ethyl carbamate is metabolised via three main metabolic pathways, namely, hydrolysis, hydroxylation and side-chain oxidation. Vinyl carbamate epoxide, one of the metabolic intermediates of EC via side-chain oxidation, is considered the main metabolite responsible for its carcinogenicity. It binds covalently to nucleic acids and proteins, resulting in the formation of adducts including those that have been shown to induce base-pair substitutions in DNA from tumour tissue. Vinyl carbamate epoxide is further metabolised to carbon dioxide and ammonia and excreted in the urine.

Developmental Toxicity

While no multigeneration studies that meet currently accepted standard protocols are available, dose-related increases in skeletal anomalies have been observed in mice given single oral doses of EC at 300 – 1000 mg/kg bw on day 11 of gestation. Increases in external malformations and skeletal abnormalities have also been reported in rats given EC at daily doses of 1000 mg/kg bw by gavage for 1, 2, 6 or 7 consecutive days during the period of organogenesis (days 6 – 13 of gestation).

Treatment of female mice with single or multiple doses of EC during gestation or lactation was found to increase the incidence or multiplicity of tumours in the adult offspring compared with untreated controls.

Mutagenicity and Genotoxicity

Ethyl carbamate is considered a genotoxic mutagen in experimental animals. It has been tested in a large number of genotoxicity studies in vitro and in vivo. The results of assays for point mutations were uniformly negative for mouse lymphoma cells, while assays in bacterial, yeast and other types of mammalian cells produced variable results. Results of assays in somatic cells in vivo (including tests for induction of chromosomal aberrations, micronucleus formation and sister chromatid exchange) were almost uniformly positive. The assay for micronucleus formation in mice showed the strongest positive response, and co-administration of ethanol delayed, but did not inhibit the genotoxicity of EC in this assay. The results of assays for dominant lethal mutations or specific locus tests in mice, given EC either by intraperitoneal injection or in drinking-water, revealed no evidence for genotoxicity in mammalian germ cells in vivo.

Observation in Human

There are at present very few EC data in humans. JECFA considered in its 2005 evaluation that the human data available were not of a quality that could be used for hazard characterisation.Risk assessment for ethyl carbamate.

JECFA evaluated EC in its 64th meeting in 2005 and concluded that EC was genotoxic and a multisite carcinogen in all animal species tested and a potential carcinogen in human.

The pivotal toxicological study for risk assessment was a carcinogenicity study of EC in mice. The increased incidence of lung tumour (alveolar and bronchiolar adenoma or carcinoma) was considered the critical toxicological end-point of concern. The associated dose-response data were analysed and the Benchmark Dose Lower Confidence Limit 10% (BMDL10) for EC was determined.

The BMDL10 represents the lower bound of a 95% confidence interval on a benchmark dose corresponding to a 10% tumour incidence (i.e. 10% extra risk of tumours) in experimental animals. The EC margin of exposure (MOE), defined as the ratio of BMDL10 to the estimated dietary intake of EC in humans, was then calculated and used by the Committee as an index for the assessment of the potential health risk posed by EC to humans.

When the estimated intake of EC in foods excluding alcoholic beverages 14 (15 ng/kg bw/day) was compared with the lower end (0.3 mg/kg bw/day) of the range of BMDL10 values (0.3 to 0.5 mg/kg bw/day) for alveolar / bronchiolar neoplasms in mice, a MOE of 20,000 was estimated. With the inclusion of alcoholic beverages in the estimated intake of EC (80 ng/kg bw/day), the resulting MOE was only 3,800. The Committee concluded that exposure to EC in foods other than alcoholic beverages would be of low concern, but the estimated health risk posed by total dietary exposure to EC (in all foods and beverages) would be of potential concern and therefore mitigation measures to reduce concentrations of EC in some alcoholic beverages should be continued.

Fermentation Microorganisms and Flavor Changes in Fermented Foods

Lactic acid bacteria influence the flavor of fermented foods in a variety of ways. In many cases, the most obvious change in a lactic acid fermentation is the production of acid and lowering pH thatresults in an increase in sourness. Since most of the acid produced in fermentations will be produced by the metabolism of sugars, sweetness will likely decrease as sourness increases. The production of volatile flavor components tends to be the first mechanism considered for the development of flavor specific to a particular fermented food. In addition to this direct mechanism, however, there are less direct ways in which fermentation microorganisms affect flavor. Lowering the pH in lactic acid fermentations may reduce the activity or completely inactivate enzymes in the plant that generate either flavor components or flavor precursor compounds.

Finally, the fermentation microorganisms may directly metabolize precursor flavor compounds or flavor components themselves.

Prevention of Flavor Formation

Purge-and-trap analysis of the volatile components found in cucumber slurries before and after cucumbers were fermented in a 2% reduced-salt brine. Comparison of volatile components before and after fermentation led to the conclusion that the major effect of the fermentation on flavor volatiles was to prevent enzymatic formation of E,Z-2,6-nonadienal and 2-nonenal by enzymes present in cucumbers (Zhou and McFeeters 1998). These aldehydes are the major compounds responsible for fresh cucumber flavor (Schieberle and others 1990). However, a few days into cucumber fermentation, the pH drops low enough to inactivate the enzymes which form these compounds when cucumber tissue is disrupted. Among the volatile components identified in the fermented cucumbers, only benzaldehyde, ethyl benzene, and o-xylene were not observed in fresh cucumber slurries. The lack of the flavor impact of volatile aldehydes is certainly the major effect of the fermentation on flavor.

More recently, Marsili and Miller (2000) found a low volatility flavor impact compound in fermented pickled cucumber brines. Addition of saturating salt to brine samples heated to 50 °C, sampling with an SPME (solid-phase microextraction) fiber and followed by GC-olfactometry led to recognition of a compound with an odor close to that of the fermentation brine. The compound with a fermentation brine odor was identified as trans-4-hexenoic acid. They also tentatively identified the presence of cis-4-hexenoic acid. In a reconstitution experiment, a solution that contained 25 ppm trans- 4-hexenoic acid, 10 ppm phenyl ethyl alcohol, 0.65% lactic acid, 0.05% acetic acid, and 8% NaCl had an odor very similar to that of brine from fermented cucumbers. The concentrations of the lactic acid, acetic acid, and NaCl are reasonable for commercial brines after the completion of fermentation. The addition of phenyl ethyl alcohol gave only a small improvement in the odor match. Thus, the trans-4-hexenoic acid was the key component in the simulated brine solution. Unfortunately, the origin of trans-4-hexenoic acid in fermentation brines is not known.

Volatile odor changes caused by lactic acid fermentation

In contrast to the cucumber fermentation, where most volatile components did not change substantially, Czerny and Schieberle (2002) observed a number of changes in odorants when whole meal wheat flour was fermented by a commercial sourdough starter culture. Using aroma extract dilution analysis, they identified 14 compounds that had substantial odor intensity in fermented sourdough. All of these odorants were present in dough before and after fermentation, so the fermentation did not result in formation of compounds not present in the flour, nor did fermentation completely remove odor components. However, most of the compounds changed during fermentation, with changes ranging from 7-fold decreases to 9-fold increases in concentration.

Flavor precursor formation caused by lactic acid fermentation

When sourdough bread is baked, one of the characteristic odor compounds generated in the baking process is 2-acetyl pyrroline, which has a roasted, popcorn-like odor, with an odor threshold of only 20 ppt (Schieberle 1995). The precursor of 2-acetyl pyrroline during baking is ornithine. Ornithine is synthesized from free arginine by heterofermentative lactic acid bacteria. De Angelis and others (2002) have shown that Lactobacillus sanfranciscensis has the enzymes arginine deiminase and ornithine transcarbamoylase, which can convert arginine to ornithine. However, L. sanfranciscen sis only increased the ornithine concentrations during dough fermentation when additional arginine was added. Thiele and Vogel (2002) reported that ornithine accumulated during sourdough fermentation only when Lactobacillus pontis was used as the starter culture. No increase in ornithine was found when the starter was L. sanfranciscensis. This result indicates the difficulty that can be encountered between showing that a fermentative organism can carry out a reaction and showing that, in the complex conditions existing in a food fermentation, the expected transformation will, in fact, occur.

While 2-acetyl-pyrroline is apparently formed from ornithine by chemical reactions in bread baking, it may be formed by lactic acid bacteria in wine. It is one of the compounds associated with the highly objectionable ‘mousy’ off-flavor in wine. Costello and Henschke (2002) showed that Lactobacillus hilgardii, a lactic acid bacterium that occurs in wine, could produce the compound from ornithine in a defined medium with a high cell density. Formation of 2-acetyl-pyrroline also required fructose as a fermentable sugar, ethanol, and iron (Fe^{+2}) in the medium.

BIBLIOGRAPHY

- Alverson, D. L., Freeberg, M. H., Murawski, S. A., and Pope, J.G. 1994. A global assessment of fisheries bycatch and discards. FAO Fisheries Technical Paper, 339. FAO, Rome. 233pp.
- Baird, RW and Guenther, TJ (1995). Account of harbour porpoise (Phocoena phocoena) strandings and bycatches along the coast of British Columbia. Rep. Int. Whal. Commn. Special Issue. 16: 159-168.
- Barlow, J & Cameron, GA (2003). Field experiments show that acoustic pingers reduce marine mammal bycatch in the Californian drift gill net fishery. Marine Mammal Science. 19: 265-283.
- Bell J.D., Ratner B.D., Stobutzki I., Oliver J. Addressing the coral reef crisis in developing countries (2006) Ocean and Coastal Management, 49 (12), pp. 976-985.
- Caswell, H, Brault, S, Read, AJ, and TD Smith (1998). Harbour porpoise and fisheries: an uncertain analysis of incidental mortality. Ecological Applications. 8(4): 1226-1238.
- Charles, A. T. 2001. Sustainable Fishery Systems. Blackwell Science, London. 384 pp.
- Cochrane, K.L. 2000. Reconciling sustainability, economic efficiency and equity in fisheries: the one that got away? Fish and Fisheries, 1: 3-21.
- Cooper A. B. (2006) A Guide to Fisheries Stock Assessment From Data to Recommendations" University of New Hampshire.
- Davis K G and Quelch G D (2008) Observer Professionalism Working Group (OPWG) 5th International Fisheries Observer Conference (IFOC) Report. 87pp.
- de Vries, Gerda; Hillen, Thomas; Lewis, Mark; Schonfisch, Birgitt and Muller, Johannes (2006) A Course in Mathematical Biology SIAM. ISBN 978-0-89871-612-2
- Demaster, DJ, Fowler, CW, Perry, SL, and ME Richlen (2001). Predation and competition: the impact of fisheries on marine mammal populations over the next one hundred years. Journal of Mammology. 82: 641-651.
- Donovan, GP. 1994. Developments on issues relating to the incidental catches of cetaceans since 1992 and the UNCED conference. Report of the IWC

(Special Issue). 15: 609-613.

- Encyclopædia Britannica Online 25 August 2005 "Population Ecology" article section on Population Cycles
- European Commission. 1994. The New Common Fisheries Policy. Office for Official Publications of the European Communities, Luxembourg. 46pp.
- FAO. 1993. Report of the Expert Consultation on the regulation of fishing effort (fishing mortality). A preparatory meeting for the FAO World Conference on fisheries management and development. Rome, 17-26 January 1983. FAO Fisheries Report R289. FAO, Rome. 34pp.
- FAO. 1995b. Precautionary Approach to Fisheries. Part 1: Guidelines on the Precautionary Approach to Capture Fisheries and Species Introductions. FAO Fisheries Technical Paper 350/1. FAO, Rome. 52pp.
- FAO. 1996. FAO Technical Guidelines for Responsible Fisheries No. 2. Precautionary Approach to Capture Fisheries. FAO, Rome. 54pp.
- FAO. 1997. FAO Technical Guidelines for Responsible Fisheries No. 4: Fisheries Management. FAO, Rome. 82pp.
- FAO. 2000. The State of World Fisheries and Aquaculture. 2000. FAO, Rome.
- Fernø, A. and Olsen, S. (eds.). 1994. Marine Fish Behaviour in Capture and Abundance Estimation. Fishing News Books, Blackwell Science Ltd., Oxford. 221pp.
- Fishing Technology Service (2008) Managing bycatch and reducing discards: Taking it to the next level
- Fogarty, M.J. & Murawski S.A. 1998. Large-scale disturbance and the structure of marine ecosystems: fishery impacts on Georges Bank. Ecological Applications, 8. S6-S22.
- Gabriel, O. (ed.). (in prep.). Fish Catching Methods of the World. (Fourth Edition). Blackwell Science Ltd., Oxford. 448 pp.
- Gearin, PJ, Gosho, ME, Laake, JL, Cooke, L, DeLong, R, and KM Hughes (2000). Experimental testing of acoustic alarms (pingers) to reduce bycatch of harbor porpoise, Phocoena phocoena, in the state of Washington. Journal of Cetacean Research and Management. 2:1-9.
- Geritz SA and Kisdi E (2004). On the mechanistic underpinning of discrete-time population models with complex dynamics. J Theor Biol. 2004 May 21;228(2):261-9.

Index